Roswell Park

Epitome of the History of Medicine

Based upon a Course of Lectures Delivered in the University of Buffalo

Roswell Park

Epitome of the History of Medicine
Based upon a Course of Lectures Delivered in the University of Buffalo

ISBN/EAN: 9783337779054

Printed in Europe, USA, Canada, Australia, Japan

Cover: Foto ©berggeist007 / pixelio.de

More available books at **www.hansebooks.com**

AN EPITOME

OF THE

HISTORY OF MEDICINE.

BY

ROSWELL PARK, A.M., M.D.,

PROFESSOR OF SURGERY IN THE MEDICAL DEPARTMENT OF THE UNIVERSITY OF BUFFALO, ETC.

BASED UPON A COURSE OF LECTURES DELIVERED IN THE
UNIVERSITY OF BUFFALO.

Illustrated with Portraits and Other Engravings.

PHILADELPHIA, NEW YORK, CHICAGO:
THE F. A. DAVIS COMPANY, PUBLISHERS.
1898.

COPYRIGHT, 1897,
BY
THE F. A. DAVIS COMPANY.

[Registered at Stationers' Hall, London, Eng.]

Philadelphia, Pa., U. S. A.:
The Medical Bulletin Printing-House,
1916 Cherry Street.

"Destiny Reserves for us Repose Enough." —FERNEL.

TO MY COLLEAGUES

IN THE

MEDICAL FACULTY OF THE UNIVERSITY OF BUFFALO,

WHO AUTHORIZED AND ENCOURAGED THIS FIRST ATTEMPT IN THE MEDICAL SCHOOLS OF THIS COUNTRY TO GIVE SYSTEMATIC INSTRUCTION IN THE HISTORY OF THE SCIENCE WHICH THEY TEACH,

THIS BOOK

IS DEDICATED.

PREFACE.

THE history of medicine has been sadly neglected in our medical schools. The valuable and fruitful lessons which it tells of *what not to do* have been completely disregarded, and in consequence the same gross errors have over and over been repeated. The following pages represent an effort to bring the most important facts and events comprised within such history into the compass of a medical curriculum, and, at the same time, to rehearse them in such manner that the book may be useful and acceptable to the interested layman,—*i.e.*, to popularize the subject. This effort first took form in a series of lectures given in the Medical Department of the University of Buffalo. The subject-matter of these lectures has been rearranged, enlarged, and edited, in order to make it more presentable for easy reading and reference. I have also tried, so far as I could in such brief space, to indicate the relationship which has ever existed between medicine, philosophy, natural science, theology, and even *belles-lettres*. Particularly is the history of medicine inseparable from a consideration of the various notions and beliefs that have at times shaken the very foundation of Christendom and the Church, and for reasons which appear throughout the book.

The history of medicine is really a history of human error and of human discovery. During the past two thousand years it is hard to say which has prevailed. Notwithstanding, had it not been for the latter the total of the former would have been vastly greater. A large part of my effort has been devoted to considering the causes which conspired to prevent the more rapid development of our art. If among these the frowning or forbidding attitude of the Church figures most prominently, it must not be regarded as any expression of a quarrel with the Church of to-day. But let any one interested read President White's *History of the Warfare of Science with Theology*, the best presentation of the subject, and he can take no issue with my statements.

Reverence for the true, the beautiful, and the good has characterized physicians in all times and climes. But little of the true, the beautiful, or the good crept into the transactions of the Church for many centuries, and we suffer, to-day, more from its interference in time past than from all other causes combined. The same may be said of theology, which is as separate from religion as darkness from light. Only when students of science emancipated themselves from the prejudices and superstitions of the theologians did medicine make more than barely perceptible progress.

In this connection I would like to quote a paragraph from an article by King, in the *Nineteenth Century* for 1893: "The difficulties under which medical science labored may be estimated from the fact that dissection was forbidden by the clergy of the Middle Ages on the ground that it was impious to mutilate a form made in the image of God. We do not find this pious objection interfering with such mutilation when effected by means of the rack and wheel and such other clerical, rather than medical, instruments."

Written history is, to a certain extent at least, plagiarism; and I make no apology for having borrowed my facts from whatever source could best furnish them, but wish cheerfully and publicly to acknowledge my indebtedness to the works mentioned below, those especially of Renouard, Baas, and Sprengel, and to various biographical dictionaries. I have not even scrupled to take bodily sentences or expressions from these authorities, but have tried to so indicate them when I could.

The writer takes pleasure in acknowledging here the obligations which both he and the publishers feel to Dr. Joseph H. Hunt, of Brooklyn, N. Y., from whose extensive and valuable collection have been furnished the originals for most of the portraits in the following pages, and to Dr. F. P. Henry, Honorary Librarian of the College of Physicians of Philadelphia, through whose courtesy was obtained the privilege of reproducing the illustrations of instruments and operations from some of the rare old works in the college library. The kind co-operation of these gentlemen has given a distinct and added value to the contents of this little work.

LIST OF PRINCIPAL WORKS CONSULTED.

BAAS, *Outlines of the History of Medicine.* Translated by Henderson. New York, 1889.
BERDOE, *Origin and Growth of the Healing Art.* London, 1893.
BOUCHUT, *Histoire de la Médecine.* Paris, 1873.
DEZEIMERIS. *Lettres sur l'Histoire de la Médecine.* Paris, 1838.
Dictionnaire Historique de la Médecine. Paris, 1828.
HAESER, *Geschichte der Medicin.* Jena, 1853.
HIRSCH, *Biographisches Lexikon des Hervorragenden der Aerzte aller Zeiten und Völker.* Wien und Leipzig, 1884.
PORTAL, *Histoire de l'Anatomie et de la Chirurgie.* Paris, 1770.
SOUTH, *Memorials of the Craft of Surgery in England.* London, 1886.
SPRENGEL, *Geschichte der Chirurgie.* Halle, 1819.

CONTENTS.

CHAPTER I.

PAGE

Medicine Among the Hebrews, the Egyptians, the Orientals, the Chinese, and the Early Greeks.—The Asclepiadæ.— Further Arrangement into Periods (Renouard's Classification). THE AGE OF FOUNDATION.—The Primitive ; Sacred, or Mystic ; and Philosophic Periods. —Systems in Vogue : Dogmatism, Methodism, Empiricism, Eclecticism.—Hippocrates. 1–29

CHAPTER II.

AGE OF FOUNDATION (*continued*).—*Anatomic Period :* Influence of the Alexandrian Library. Herophilus and Erasistratus. Aretæus. Celsus. Galen. — *Empiricism :* Asclepiades.— *Methodism :* Themison. —*Eclecticism.* AGE OF TRANSITION.—*Greek Period :* Oribasius. Ætius. Alexander of Tralles. Paulus Ægineta. 30–56

CHAPTER III.

AGE OF TRANSITION (*continued*).—*Arabic Period :* Alkindus. Mesue. Rhazes. Haly-Abbas. Avicenna. Albucassis. Avenzoar. Averroës. Maimonides.—*School of Salernum :* Constantinus Africanus. Roger of Salerno. Roland of Parma. The Four Masters. John of Procida. 57–85

CHAPTER IV

AGE OF TRANSITION (*concluded*).—*The School of Montpellier :* Raimond Lulli. John of Gaddesden. Arnold of Villanova. Establishment of Various Universities. Gerard of Cremona. William of Salicet. Lanfranc. Mondino. Guy de Chauliac. AGE OF RENOVATION.— *Erudite Period*, including the Fifteenth and Sixteenth Centuries. Thomas Linacre. Sylvius. Vesalius. Columbus. Eustachius. Fallopius. Fabricius ab Aquapendente. Fabricius Hildanus. 86–113

CHAPTER V.

AGE OF RENOVATION (*continued*).—*Erudite Period* (*continued*) *:* Benivieni. Jean Fernel. Porta. Severino. Incorporation of Brother-

(ix)

hood of St. Côme into the University of Paris. Ambroise Paré. Guillemeau. *Influence of the Occult Sciences:* Agrippa. Jerome Cardan. Paracelsus. Botal. Joubert. 114–147

CHAPTER VI.

AGE OF RENOVATION (*continued*).—Student-life During the Fifteenth and Sixteenth Centuries. Ceremonials Previous to Dissection.—*Reform Period:* The Seventeenth, Eighteenth, and Nineteenth Centuries. Modern Realism in Medicine and Science. Introduction of the Cell-doctrine. Discovery of the Circulation. William Harvey. Malpighi. Leuwenhoek. Correct Doctrine of Respiration. Discovery of the Lymphatic Circulation. The Nervous System. Discovery of Cinchona. Development in Obstetric Art, in Medical Jurisprudence, and in Oral Clinical Teaching. Van Helmont.—*The Iatrochemical System:* Le Böe. Thomas Willis. 148–170

CHAPTER VII.

AGE OF RENOVATION (*continued*).—*Iatromechanical School:* Santoro. Borelli. Sydenham. Sir Thomas Browne.—*Surgery:* Denis. F. Collot. Dionis. Baulot (Frère Jacques). Scultetus. Rau. Wiseman. Cowper. Sir C. Wren the Discoverer of Hypodermatic Medication. Anatomical Discoveries. General Condition of the Profession During the Seventeenth Century. The Eighteenth Century. Boerhaave. Gaub.—*Animism:* Stahl.—*Mechanico-dynamic System;* Hoffmann. Cullen.—*Old Vienna School:* Van Swieten. De Haën.—*Vitalism:* Bordeu. Erasmus Darwin. 171–202

CHAPTER VIII.

AGE OF RENOVATION (*continued*).—*Animal Magnetism:* Mesmer. Braid. —*Brunonianism:* John Brown.—*Realism:* Pinel. Bichat. Avenbrugger. Werlhof. Frank.—*Surgery:* Petit. Desault. Scarpa. Gimbernat. Heister. Von Siebold. Richter. Cheselden. Monro (1st). Pott. John Hunter. B. Bell, J. Bell, C. Bell. Smellie. Denman.—*Revival of Experimental Study:* Haller. Winslöw. Portal. Vicq d'Azyr. Morgagni.—*Inoculation against Small-pox:* Lady Montagu. Edward Jenner. 203–229

CHAPTER IX.

AGE OF RENOVATION (*continued*).—The Eighteenth Century; General Considerations. Foundation of Learned Societies, etc. The Royal College of Surgeons; the Josephinum.—The Nineteenth Century Realistic Reaction Against Previous Idealism. Influence of Comte, of Claude Bernard, and of Charles Darwin. Influence Exerted by Other Sciences.—*Theory of Excitement:* Roeschlaub.

—*Stimolo and Contrastimolo:* Rasori.—*Homœopathy:* Hahnemann.
—*Isopathy, Electrohomœopathy* of Mattei.—*Cranioscopy, or Phrenology:* Gall and Spurzheim.—*The Physiological Theory:* Broussais.
—*Paris Pathological School:* Cruveilhier. Andral. Louis. Magendie. Trousseau. Claude Bernard.—*British Medicine:* Bell and Hall. Travers.—*Germany, School of Natural Philosophy:* Johannes Müller.—*School of Natural History:* Schönlein.—*New Vienna School:* Rokitansky. Skoda. 230–252

CHAPTER X.

AGE OF TRANSITION (*concluded*).—*New Vienna School (concluded):* von Hebra. Czermak and Türck. Jäger. Arlt. Gruber. Politzer.—*German School of Physiological Medicine:* Roser.—*School of Rational Medicine:* Henle.—*Pseudoparacelsism:* Rademacher.—*Hydrotherapeutics:* Priessnitz.—*Modern Vitalism:* Virchow.—*Seminalism:* Bouchut.—*Parasitism and the Germ-theory:* Davaine. Pasteur. Chauveau. Klebs. F. J. Cohn. Koch. Lister.—*Advances in Physical Diagnosis:* Laënnec. Piorry.—SURGERY: Delpech. Stromeyer. Sims. Bozeman. McDowell. Boyer. Larrey. Dupuytren. Cloquet. Civiale. Vidal. Velpeau. Malgaigne. Nélaton. Sir Astley Cooper. Brodie. Guthrie. Syme. Simpson. Langenbeck. Billroth. 253–275

CHAPTER XI.

HISTORY OF MEDICINE IN AMERICA.—The Colonial Physicians. Medical Study under Preceptors. Inoculation against Small-pox. Military Surgery During the Revolutionary War. Earliest Medical Teaching and Teachers in this Country. The First Medical Schools. Benjamin Rush. The First Medical Journals. Brief List of the Best-Known American Physicians and Surgeons. . 276–299

CHAPTER XII.

THE HISTORY OF ANÆSTHESIA.—Anæsthesia and Analgesia. Drugs Possessing Narcotic Properties in use since Prehistoric Times. Mandragora; Hemp; Hasheesh. Sulphuric Ether and the Men Concerned in its Introduction as an Anæsthetic—Long, Jackson, Wells, and Morton. Morton's First Public Demonstration of the Value of Ether. Morton Entitled to the Credit of its Introduction. Chloroform and Sir James Simpson. Cocaine and Karl Koller. 300–315

CHAPTER XIII.

THE HISTORY OF ANTISEPSIS.—Sepsis, Asepsis, and Antisepsis. The Germ-theory of Disease. Gay-Lussac's Researches. Schwann. Tyndall. Pasteur. Davaine. Lord Lister and his Epoch-making

Revolution in Surgical Methods. Modifications of his Earlier Technique without Change in Underlying Principles, which Still Remain Unshaken. Changes Effected in Consequence. Comparison of Old and Modern Statistics. 316–329

CHAPTER XIV.

AN EPITOME OF THE HISTORY OF DENTISTRY.—Rude Dentistry of Prehistoric Times. Early Instruments for Extraction Made of Lead. Dentistry on the Same Low Plane as Medicine During the First Half of the Christian Era. Dentistry Taught at the School of Salernum. Progress of the Art on the Continent. Prosthesis and Substitutes for Human Teeth. Introduction of Porcelain for Artificial Teeth; of Metal and of Vulcanized Rubber for Plates; of Plaster for Impressions. From being a Trade, Dentistry is now a Profession, in which Americans lead the World. Statistics. 330–341

LIST OF ILLUSTRATIONS.

FIG.		PAGE
1.	Æsculapius,	7
2.	Offering to Æsculapius,	9
3.	Hippocrates,	19
4.	Aulus Cornelius Celsus,	35
5.	The Conversion of Galen,	37
6.	Averroës,	64
7.	Andreas Vesalius,	105
8.	Title-page of *The Seven Books of the Anatomy of the Human Body*, by Andreas Vesalius, of Brussels, Physician to the Invincible Emperor Charles V,	106
9.	IV, Forceps for Extracting Balls. V, A Denticulated Form of Forceps	108
10.	Gabriel Fallopius,	109
11.	Forms of Forceps for Enlarging Wounds,	111
12.	Body Showing Various Kinds of Wounds,	117
13.	Mode of Extracting Leaden Bullets,	121
14.	Ambroise Paré,	124
15.	Pliers, Iron for Actual Cautery, and Seton-needles,	126
16.	Swan's Beak, Used for Dilating the Track of a Wound and Extracting a Foreign Body,	129
17.	Various Instruments for the Extraction of Balls,	131
18.	Speculums for the Mouth and Womb, etc.,	133
19.	Amputation Instruments,	135
20.	Different Forms of Trephines and Pliers,	137
21.	Philip Theophrastus Paracelsus,	143
22.	William Harvey, M.D.,	156
23.	Thomas Sydenham,	173
24.	Straight Saws and Divers Scraping Tools, Wherewith the Skull, Being Rotten or Having a Fissure, is Scraped Away,	179
25.	Surgical Treatment of Certain Dislocations,	181
26.	Various Operations on the Arms and Lower Limbs,	185
27.	Surgical Operations on the Breast, etc.,	187
28.	Boerhaave,	193
29.	John Brown, M.D.,	205
30.	Ph. Pinel,	207
31.	Marie François Xavier Bichat, M.D.,	208
32.	William Hunter, M.D., F.R.S.,	217
33.	John Hunter,	218
34.	J. F. Blumenbach,	223
35.	Edward Jenner, M.D.,	227
36.	Samuel Hahnemann,	242
37.	Rudolph Virchow,	257
38.	Bernhard von Langenbeck,	265
39.	Theodor Billroth,	266

(xiii)

FIG.		PAGE
40.	Sir Astley Cooper, Bart.,	272
41.	Sir Benjamin Collins Brodie, Bart., F.R.S.,	273
42.	B. Waterhouse, M.D.,	280
43.	Surgeon's Hall, next to Philadelphia Dispensary, Fifth Street below Library Street,	281
44.	Benjamin Rush, M.D.,	284
45.	George B. Wood, M.D.,	286
46.	Robley Dunglison, M.D.,	287
47.	Austin Flint, M.D.,	288
48.	Isaac Ray, M.D.,	289
49.	Philip Syng Physick, M.D.,	291
50.	Ephraim McDowell, M.D.,	292
51.	S. D. Gross, M.D., LL.D.,	294
52.	J. Marion Sims, M.D.,	296
53.	D. Hayes Agnew, M.D., LL.D.,	297
54.	William T. G. Morton, M.D.,	307
55.	Dr. Morton Making the First Public Demonstration of Etherization at the Massachusetts General Hospital Surrounded by the Medical Staff of that Institution, October 16, 1846,	309
56.	Lord Lister, M.D., D.C.L., LL.D., P.R.S.,	323

An Epitome of the History of Medicine.

CHAPTER I.

Medicine Among the Hebrews, the Egyptians, the Orientals, the Chinese, and the Early Greeks.—The Asclepiadæ.—Further Arrangement into Periods (Renouard's Classification).

THE AGE OF FOUNDATION.—The Primitive; Sacred, or Mystic; and Philosophic Periods.—Systems in Vogue: Dogmatism, Methodism, Empiricism, Eclecticism.—Hippocrates, born 460 B.C.

OF the origin of medicine but little need be said by way of preface, save that it must have been nearly contemporaneous with the origin of civilization. The lower animals when sick or wounded instinctively lessen or alter their diet, seek seclusion and rest, and even in certain cases seek out some particular herb or healing substance. Thus, too, does the savage in his primitive state; and experience and superstition together have led nearly all the savage tribes into certain habits and forms in case of injury or disease. For us the history of medicine must necessarily begin with the written history of events, and its earliest endeavors need detain us but a very short time. Its earliest period is enveloped in profound obscurity, and so mingled with myth and fable as to be very uncertain. It embraces an indefinite time, during which medicine was not a science, but an undigested collection of experimental notions,— vaguely described, disfigured by tradition, and often made inutile by superstition and ignorance. The earliest records of probable authenticity are perhaps to be met with in the Scriptures, from which may be gathered here and there a fair notion of Egyptian knowledge and practice. Thus we read that Joseph commanded his servants and physicians to embalm him, this being about 1700 B.C. It shows that Egypt at that time possessed a class of men who practiced the healing art, and that they also embalmed the dead, which must have both required and furnished a crude idea of general anatomy. We are also informed from other sources that so superstitious were the Egyptians that they not only scoffed at, but would stone, the embalmers, for whom they had sent, after the completion of their task. The probably mythical being whom the Egyptians called *Thoth*, whom the Greeks named *Hermes* and the Latins *Mercury*, passed among the Egyptians as the inventor of all sciences and arts. To him are attributed an enormous number of writings concerning all subjects. Some have considered him as identical with Bacchus, Zoroaster, Osiris, Isis, Serapis, Apollo, and even Shem, the son of Noah. Others have thought him to be a god. It is now almost certain that the books attributed to Hermes were not the work of any one hand or of any one age. The last six volumes of the forty-two composing the encyclopædia, with which Hermes is credited, refer to medicine, and embrace a body of doctrines fairly complete and well arranged. Of these six, the first treats of anatomy; the second, of diseases; the third, of instruments; the fourth, of remedies; the fifth, of diseases of the eye; and the sixth, of diseases of women. In completeness and arrangement it rivals, if not surpasses, the Hippocratic collection, which it antedated by perhaps a thousand years. The Egyptians appear at first to have exposed their sick in public (at least, so says Strabo), so that if any of those who passed by had been similarly attacked they might give their advice for the benefit of the sufferers. In fact, according to Herodotus, the same custom prevailed among the Babylonians and Lusitanians. At a later date all who were thus cured were required to go to the temples and there inscribe their symptoms and what had helped them. The temples of Canopus and Vulcan at Memphis became the principal depots for these records, which were kept as carefully as were the archives of the nation, and were open for public reference. These records, being under the control of the priests, were mainly studied by them, who later collected a great mass of facts of more or less importance, and endeavored to found upon the knowledge thus collected an exclusive practice of the art of medicine. In this way they formed their medical code, which was called by Diodorus the Hiera Sacra, *Sacred Book*, from whose directions they were never allowed to swerve. It was perhaps this code which was later attributed to Hermes, and that made up the collection spoken of by Clement of Alexandria. If in following these rules they could not save their patients they were held blameless, but were punished with death if any departure from them were not followed by success.

I have spoken of embalming as practiced by the Egyptians. It was of three grades: the first reserved for men of position and means, which cost one talent, and according to which the brain was removed by an opening through the nasal fossæ, and the intestines through an opening on the left side of the abdomen, after which both cavities were stuffed with spices and aromatics; then the body was washed and spread over with gum and wrapped in bandages of linen. The second grade was adopted by families of moderate means; and the third was resorted to by the poor, consisting simply in the washing of the body and maceration in lye for seventy days.

Pliny assures us that the kings of Egypt permitted the opening of corpses for the purpose of discovering the causes of disease, but this was only permitted by the Ptolemies, under whose reign anatomy was carried to a very high degree of cultivation.

The medicine of the Hebrews is known generally through the Sacred Scriptures, especially through the writings attributed to Moses, which embraced rules of the highest sagacity, especially in public hygiene. The book of

Leviticus is largely made up of rules concerning matters of public health. In the eleventh chapter, for instance, meat of the rabbit and the hog is proscribed, as apparently injurious in the climate of Egypt and India; it, however, has been suggested that there was such variation of names or interpretation thereof as to make it possible that our rabbit and hog are not the animals alluded to by Moses. The twelfth and fifteenth chapters of the same book were designed to regulate the relation of man and wife and the purification of women, their outlines being still observed in some localities by certain sects, while the hygienic measure of circumcision then insisted upon is still observed as a religious rite among the descendants of Moses. For the prevention of the spread of leprosy, the measures suggested by Moses could not now be surpassed, although ancient authors have confounded under this name divers affections, probably including syphilis, to which, however, the same hygienic rules should apply. Next to Moses in medical lore should be mentioned Solomon, to whom is attributed a very high degree of knowledge of natural history, and who, Josephus claimed, had such perfect knowledge of the properties of all the productions of nature that he availed himself of it to compound remedies extremely useful, some of which had even the virtues necessary to cast out devils.

The most conspicuous feature in the life of the Indian races is their division into castes, of which the most noble is that of the priests, or Brahmins, who in ancient times alone had the privilege of practicing medicine. Their Organon of Medicine, or collection of medical knowledge, was a book which they called *Vagadasastir*. It was not systematically arranged, and in it demonology played a large rôle. They held the human body to consist of 100,000 parts, of which 17,000 were vessels, each one of which was composed of seven tubes, giving passage to ten species of gases, which by their conflicts engendered a number of diseases. They placed the origin of the pulse in a reservoir located behind the umbilicus. This was four fingers wide by two long, and divided into 72,000 canals, distributed to all parts of the body. The physician examined not only the pulse of his patient, but the dejecta, consulted the stars, the flight of birds, noted any incidental occurrence during his visits, and made up his prognosis from a multitude of varying circumstances, omitting only those which were really valuable, namely, the symptoms indicating the state of the organs. Ancient Hindoo charlatan priests let fall from the end of a straw a drop of oil into the patient's water. If the oil was precipitated and attached itself to the bottom of the vessel, they predicted an unfavorable result; if, on the contrary, it floated, they gave a favorable prognosis. This is, so far as we know, the earliest recorded way of testing the specific gravity of the urine.

With all their absurdities, however, the Indians appear to have done some things that we scarcely do to-day: they are said to have had an ointment that caused the cicatrices of variola to disappear, and they cured the bites of venomous serpents with remedies whose composition has been lost.

The antiquity of the Chinese is simply lost in tradition and fable. From time immemorial their rulers have taken extraordinary care to prevent contact and interchange of ideas with foreigners. For 4000 years their manners, laws, religious beliefs, language, and territory have scarcely changed. In this respect they stand alone among the nations of the earth. They attribute the invention of medicine to one of their emperors named Hoâm-ti, who was the third of the first dynasty, and whose supposititious date is 2687 B.C. He is considered to be the author of the work which still serves them as a medical guide. It is, however, more probably an apochryphal book. Its philosophy was of a sphygmic kind,—*i.e.*, based upon the pulse, which they divided into the supreme or celestial, the middle, and the inferior or terrestrial; by the examination of which the Chinese physician was supposed not only to show the seat of disease, but to judge of its duration and gravity. It is related that one of the ancient Chinese emperors directed the dead bodies of criminals to be opened, but this is questionable, since it is certain that they have the most profound ignorance of rudimentary anatomy, and glaring errors abound in their system. Being thus replete with errors, and possessing no anatomical knowledge, their surgery was of the most barbarous type. No one dared attempt a bloody operation; the reduction of hernia was unknown; a cataract was regarded as beyond their resources; and even venesection was never practiced. On the other hand, they employed cups, and acupuncture, fomentation, plasters of all kinds, lotions, and baths. The moxa, or red-hot button, was in constant use, and they had their magnetizers, who appear to have been convulsionists. For a long time there existed at Pekin an Imperial School of Medicine, but now there is no such organization nor any regulation for the privilege of practicing medicine or surgery since 1792. At least until lately the country and the cities were infested with quacks, who dealt out poison and death with impunity. They practiced most murderous methods in place of the principles of midwifery. Only since the civilized missionaries have penetrated into their country has there been any improvement in this condition of affairs.

It is Greece which furnishes us with the most interesting and the most significant remains of the history of medicine during antiquity, as she furnishes every other art with the same historical advantages. During the period preceding the Trojan War there is little but myth and tradition. Leclerc catalogued some thirty divinities, heroes or heroines, who were supposed to have invented or cultivated some of the branches of medicine. *Melampus* is perhaps the first of these who immortalized himself by extraordinary cures, especially on the daughters of Proetus, King of Argos. These young princesses, having taken vows of celibacy, became subjects of hysterical monomania, with delusions, during which they imagined themselves transformed into cows and roamed the forests instead of the palaces. This nervous delusion spread to and involved many other women, and became a serious matter.

FIG. 1.—ÆSCULAPIUS.
(From a steel engraving of the marble statue in the Louvre.)

Melampus, the shepherd, having observed the purgative effects upon goats of white hellebore, gave to the young women milk in which this plant had been steeped, thereby speedily effecting a cure. Scarcely less distinguished than Melampus was *Chiron*. He was mainly distinguished because he was the preceptor of *Æsculapius*, the most eminent of early Greeks in this field. By some Æsculapius was considered the son of Apollo by the nymph Coronis.

Several cities of Greece contended for the honor of his birthplace, as they did for that of Homer. That he was famous at the time of the Argonautic expedition is seen by the fact that the twins Castor and Pollux desired him to accompany the expedition as surgeon. Be his origin what it may, Æsculapius was the leading character in medicine of all the ancients, with the possible exception of Hermes among the Egyptians; in fact, some scholars consider the two identical. Temples were erected in his honor, priests were consecrated to them, and schools of instruction were there established. It is related that Pluto, god of hell, alarmed at the diminishing number of his daily arrivals, complained to Jupiter, who destroyed the audacious healer —on which account, some wit has said, " the modern children of Æsculapius abstain from performing prodigies." But the true Æsculapians, the successors of the demigod, were imitated or copied by the crowd of charlatans and quacks, calling themselves theosophs, thaumaturgs, and so on, and not alone at that date, but for generations and centuries thereafter, Paracelsus and Mesmer being fair examples of this class. The poet Pindar, who lived seven or eight hundred years after Æsculapius, says that he cured ulcers, wounds, fever, and pain of all who applied to him by enchantment, potions, incisions, and by external applications.[1]

The followers of Æsculapius, and the priests in the temples dedicated to him, soon formed a separate caste, transmitting from one to another, as a family heritage, their medical knowledge. At first no one was admitted to practice the sacred science unless he joined the priesthood, although later this secrecy was relaxed. They initiated strangers, provided they fulfilled the test which they made. Some kind of medical instruction was given in each temple. The three most celebrated temples to Æsculapius were that of Rhodes, already extinct by the time of Hippocrates;

---

[1] Third Pythian Ode.

that of Cnidus, which published a small repertory; and
finally that of Cos, most celebrated of all, because of the
illustrious men who emanated from this school. In these
temples votive tablets were fastened in large numbers, after
the fashion of the Egyptians, the same giving the name of
the patient, his affliction, and the manner of his cure. For
example, such a one as this: "Julien vomited blood, and
appeared lost beyond recovery. The oracle ordered him to

Fig. 2.—Offering to Æsculapius.
(From an engraving by George Cooke of a painting by Guérin.)

take the pine-seeds from the altar, which they had three
days mingled with honey; he did so, and was cured.
Having solemnly thanked the god, he went away." There
is reason to think that the priests of these temples made for
their own uses much more minute and accurate accounts,
which should be of some real service, since the writings
which have come down to us evince a habit of close observation and clear description of disease. During the Trojan

War two men are frequently mentioned by Homer as possessing great surgical skill. These were *Machaon* and *Podalirius*. They were regarded as sons of Æsculapius, the former being the elder. The first account of venesection, although not authentic, refers to the bleeding practiced by the latter upon the daughter of the King of Caria, upon whose shores Podalirius was cast by tempest after the ruin of Priam's kingdom. Whether he was the first of all men to practice it or not, it is certain that the act of venesection goes back long prior to the era of Hippocrates, who speaks of it as frequently performed.

Many of the deities upon Olympus seem at one time or another to have usurped medical functions. Apollo, the reputed father of Æsculapius, appropriated nearly everything under the name of Pæon, who assumed the privilege of exciting or subduing epidemics. Juno was supposed to preside at accouchements, and in both the Iliad and Odyssey it is indicated that Apollo was considered as the cause of all the natural deaths among men, and Diana of those among women.

The long Trojan War appears to have been an epoch-making event in the medical and surgical history of those times, as was the Civil War recently in our country. Certain vague and indefinite practices then took more fixed form, and from that time on medicine may be said to have been furnished with a history. After the dethronement of Priam and the destruction of his capital, navigation was free and unrestricted. The Hellenists covered with their colonies both shores of the Mediterranean, and their navigators even passed the pillars of Hercules. By these means the worship of Æsculapius passed from Greece into what is now Asia, Africa, and Italy. In his temple at Epidaurus was a statue of colossal size made of gold and ivory. The dialogues of Plato, especially the Phædo, make it apparent that the cock was the animal sacrificed to him, and hence sacred to the god of medicine. The priests attached to his worship were called *Asclepiadæ*, or descendants of Æsculapius. The temples were usually hygienically located near thermal springs or fountains and among groves. Pilgrimages were made from all quarters, and these localities became veritable health-resorts. A well-regulated dietary, pure air, temperate habits, and faith stimulated to a fanatical degree combined and sufficed for cures which even nowadays would be regarded as wonderful. The priests prescribed venesection, purgatives, emetics, friction, sea-baths, and mineral waters, as they appeared to be indicated. The imagination of the patient was continually stimulated, and at the same time controlled. Before interrogating the oracles they must be purified by abstinence, prayer, and sacrifice. Sometimes they were obliged to lie in the temple for one or more nights. The gods sometimes revealed themselves in mysterious ways, at times devouring the cakes upon the altars under the guise of a serpent, or again causing dreams which were to be interpreted by the priests. There can be no doubt that sometimes, at least, the grossest frauds and the basest trickery were relied upon for the purpose of impressing the minds of those weakened by abstinence or influenced by drugs. Mercenary considerations were not lacking; moreover, cures were often not obtained until zeal had been redoubled by largely increased contributions to the treasury of the temples. In the neighborhood of many of these temples serpents abounded, non-venomous and easily tamed. These were employed by the priests in various supernatural performances by which the ignorant people were astonished and profoundly impressed. In fact, the serpent and the serpent-myth played a very large rôle in the early history of medicine as well as that of religion and religious symbolism.

It will thus be seen that during the space of about 700 years medicine underwent a transformation in Greece. It was first domestic and popular, practiced by shepherds,

12   THE HISTORY OF MEDICINE.

soldiers, and others; then became sacerdotal; after the Trojan War it was confined to the vicinity of the temples and practiced in the name of some divinity; and finally it was wrapped in mystery and mystic symbolism, where superstition was played upon and credulity made to pay its reward. Down to the time of Hippocrates the Asclepiadæ rendered some genuine service to science, especially by inculcating habits of observation, in which Hippocrates excelled above all. Later, however, down to the time of the Christian era, medicine in the temples declined, and became, in fact, a system based upon the grossest jugglery.

It is time now that we make a systematic attempt to classify events in the history of medicine, and to recognize certain distinct epochs as they have occurred. For this purpose I know of no better arrangement than that of Renouard, which, in the main, I shall follow, at least during the forepart of this book. In this sense he divides the past into three ages, known, respectively, as the *Age of Foundation*, the *Age of Transition*, and the *Age of Renovation*. Each of these chronological divisions is subdivided into periods, of which the first contains four:—

AGE OF FOUNDATION.—1. The *Primitive Period*, or that of *Instinct*, beginning with myth, and ending with the destruction of Troy 1184 years before Christ.

2. The *Sacred*, or *Mystic*, *Period*, ending with the dispersion of the Pythagorean Society, 500 years before Christ.

3. The *Philosophic Period*, terminating with the foundation of the Alexandrian library, 320 years before Christ.

4. The *Anatomic Period*, ending with the death of Galen, about A.D. 200.

The SECOND AGE, or that of TRANSITION, is divided into a fifth, or *Greek Period*, ending at the burning of the Alexandrian library, A.D. 640, and a sixth, *Arabic Period*, ending with the revival of letters, A.D. 1400.

The THIRD AGE, or that of RENOVATION, includes the seventh, or *Erudite Period*, comprising the fifteenth and sixteenth centuries, and eighth, or *Reform Period*, comprising the seventeenth, eighteenth, and nineteenth centuries.

Examining this table for a moment, it will be seen that so far we have dealt with the Primitive Period and the Sacred, or Mystic, Period. Before passing on to the Philosophic Period let us for a moment follow Renouard, who likens the three schools of medical belief in the earlier part of the Primary Age, or the Age of Foundation, to the three schools of cosmogony, which obtained among the Greeks. The first of these was headed by Pythagoras, who regarded the universe as inhabited by acknowledged sentient principles which governed all substances in a determined way for preconceived purposes. Animals, plants, and even minerals were supposed to possess vivifying spirits, and above them all was a supreme principle. To this school corresponded the so-called *Dogmatic School* of medicine, attributed to Hippocrates, which was the precursor of modern vitalism, and regarded diseases as indivisible units from beginning to termination; in other words, they consisted of a regular programme of characteristic systems, successive periods, and of long course, either for the better or worse; that was one of the characteristic dogmas of the Hippocratic teaching. The Second System of cosmogony was that founded by Leucippus and Democritus, who explained all natural phenomena without recourse to the intervention of intelligent principles. All things for them existed as the necessary result of the eternal laws of matter. They denied preconceived purposes and ridiculed final causes. To this system corresponded that in medicine which has been termed *Methodism* (medically and literally speaking) and which recognized as its founders Æsculapius and Themison. The believers in this doctrine attempted to apply the atomic theory of Democritus and Epicurus to the theory and practice of medicine. Atoms of various size were supposed to pass and repass without cessation through cavities or pores in the human body. So long as the atoms and pores maintained a normal relationship of size and proportion health was maintained, but it was deranged so soon as the exactness of these relations was destroyed or interfered with. The Dogmatists considered vital reaction as a primary phenomenon, while with the Methodists it was secondary. The Third System of cosmogony, founded by Parmenides and Pyrrho, believed in the natural improvement of bodies in their endless reproduction and change, and concluded that wisdom consisted in remaining in doubt; in other words, they were the agnostics of that day. "What is the use," said they, "of fatiguing the mind in endeavoring to comprehend what is beyond its capability?" Later they were known as Skeptics and Zetetics, to indicate that they were always in search of truth without flattering themselves that they had found it. To them corresponded a third class of physicians, with Philinus and Serapis at their head, who deemed that proximate causes and primitive phenomena of disease were inaccessible to observation; that all that is affirmed on these subjects is purely hypothetical, and hence unworthy of consideration in choosing treatment. For them objective symptoms—or, as we would say, signs—constituted the natural history of disease, they thus believing that their remedies could only be suggested by experience, since nothing else could reveal itself to them. They therefore took the name of *Empirics*.

Finally a fourth class of physicians arose who would not adopt any one of these systems exclusively, but chose from each what seemed to them most reasonable and satisfactory. They called themselves *Eclectics*, wishing thereby to imply that they made rational choice of what seemed best. The idea conveyed in the term "eclecticism" has been fairly criticised for this reason: eclecticism is in reality neither a system nor a theory; it is individual pretension elevated to the dignity of dogma. The true eclectic recognizes no other rule than his particular taste, reason, or fancy, and two or more eclectics have little or nothing in common. If that were true two thousand years ago, it is not much less so to-day. The eclectic carefully avoids the discussion of principles, and has neither taste nor capacity for abstract reasoning, although he may be a good practitioner; not that he has no ideas, but that his ideas form no working system. With him medical tact—*i.e.*, cultivated instinct —replaces principle.

The eclectic of our day, however, is only an empiric in disguise,—that is, a man whose opinions are based on comparison of observed facts, but whose theoretical ideas do not go beyond phenomena.

In older days philosophy embraced the whole of human knowledge, and the philosopher was not permitted to be unacquainted with any of its branches. Now physics, metaphysics, natural history, etc., are arranged into separate sciences, and the sum-total of knowledge is too great to be compassed by any one man.

Pythagoras was the last of the Greek sages who made use of hieroglyphic writings and transmitted his doctrine in ancient language. Born at Samos, he was, first of all, an athlete; but one day, hearing a lecture on immortality of the soul, he was thereby so strongly attracted to philosophy that he renounced all other occupation to devote himself to it. He studied arduously in Egypt, in Phœnicia, in Chaldea, and even, it is said, in India, where he was initiated into the secrets of the Brahmins and Magi. Finally, returning to his own country, he was received by the tyrant Polycrates, but not made to feel at home. Starting on his travels again, he assisted at one of the Olympic games, and, being recognized, was warmly greeted. He sailed to the south of Italy, landed at Crotona, and lodged with Milo, the athlete. Commencing here his lectures, he soon gathered around him a great number of disciples, of whom he required a very severe novitiate, lasting even five or six years, during which they had to abstain almost entirely from conversation, and live upon a very frugal diet. Those only who persevered were initiated later into the mysteries of the order. His disciples had for him most profound veneration, and were accustomed to decide all disputes with: "The master has said it." Pythagoras possessed immense knowledge; he invented the theorem of the square of the hypothenuse, and he first divided the year into 365 days and 6 hours. He seems to have suspected the movements of our planetary system. He traveled from place to place, and founded schools and communities wherever he went, which exercised, at least at first, only the happiest influence; but the success and influence which their learning gave them later made his disciples bold, and then dishonest, and his communities were finally dispersed by angry mobs, which forced their members to conceal or expatriate themselves; and so, even during the life-time of its founder, the Pythagorean Society was destroyed, and never reconstructed.

With Pythagoras and his disciples numbers played a very important rôle, and the so-called language of numbers was first taught by him. He considered the unit as the essential principle of all things, and designated God by the figure 1 and matter by the figure 2, and then he expressed the universe by 12, as representing the juxtaposition of 1 and 2. As 12 results from multiplying 3 by 4, he conceived the universe as composed of three distinct worlds, each of which was developed in four concentric spheres, and these spheres corresponded to the primitive elements of fire, air, earth, and water. The application of the number 12 to express the universe Pythagoras had received from the Chaldeans and Egyptians—it being the origin of the institution of the zodiac. Although this is digressing, it serves to show what enormous importance the people of

Island of Cos, of a family in which the practice of medicine was hereditary, who traced their ancestors on the male side to Æsculapius, and on the female side to Hercules. The individual to whom every one refers under this name was the second of seven; the date of his birth goes back to 460 B.C., but of his life and his age at death we do not know; some say he lived to be over one hundred years of age. It is certain that he traveled widely, since his writings

FIG. 3.—HIPPOCRATES.

evince the knowledge thus gained. He was a contemporary of Socrates, although somewhat younger, and lived in the age of Pericles,—the golden age for science and art in Greece.

The Island of Cos is now called Stan-Co, and is situated not far from the coast of Ionia. Formerly it was considered as having a most salubrious climate; now that it is under the dominion of the Turks, it is considered most unhealthy. It possessed a temple dedicated to Æsculapius and a celebrated medical school. But Hippocrates, not satisfied with what he could learn here, visited the principal foreign cities, and seems to have been a most accurate and painstaking observer and collector of notes. That he achieved great renown in his life is known, since Plato and even Aristotle refer to him as their authority in very many matters. His children and grandchildren followed in his footsteps, and published their writings under the same name; it has, therefore, become difficult to distinguish his works from theirs. Finally, authors more unscrupulous, who bore no relationship to him, attached his name to their own writings. But the true were, as a rule, easily distinguished from the spurious, and were carefully separated by those in charge of the Alexandrian library.

The enumeration of his writings by different authors varies very much. Renouard, who seems to have studied the subject very carefully, gives the following as appearing to him to be the authentic list of writings of Hippocrates the Second,—*i.e.*, the Great: The *Prognostic*, the *Aphorisms*, the first and third books of *Epidemics*, that on *Regimen in Acute Diseases*, that on *Airs, Waters, and Places*, that on *Articulations and Luxations*, that on *Fractures*, and the *Mochlic*, or the treatise on instruments and reduction. This list does not comprise the fourth part of the entire Hippocratic collection, but its authenticity appears to be undoubted, and it suffices, as Renouard says, to justify the enthusiasm of his contemporaries and the admiration of posterity. Later, joined with the writings of Pythagoras, Plato, Aristotle, and others, they constituted the so-called *Hippocratic collection*, which was a definite part of the great libraries of Alexandria and Pergamos, and formed the most ancient authentic monument of medical science.

Respect for the bodies of the dead was a religious observance in all Greece, and prevented the dissection of the human body. Consequently the knowledge of anatomy possessed by Hippocrates must have been meagre. Nevertheless, he described lesions, like wounds of the head, of the heart, the glands, the nature of bones, etc. It being impossible to establish a physiology without an anatomical basis, it is not strange that we find but little physiology in the Hippocratic writings, and that this little is very crude and incorrect. Arteries and veins were confounded, and nerves, tendons, ligaments, and membranes were represented as analogous or interchangeable tissues. The physiologists of those days abandoned themselves to transcendental speculation concerning the nature and principles of life, which some placed in moisture, others in fire, etc. Speculation, thus run wild, prevented such accurate observation as might have greatly enhanced the progress of physiological knowledge.

Hippocrates wrote at least three treatises concerning hygiene: The first, on *Airs, Waters, and Places;* the second, on *Regimen;* the third, on *Salubrious Diet,*— practically an abridgement of the preceding, in which he recommends the habit of taking one or two vomits systematically every month. The classification of diseases into internal or medical, and external or surgical, is not modern, but is due to Hippocrates; neither is it philosophic, although it is very convenient.

With so little knowledge of physiology and pathology as the ancients had, it is not strange that they ascribed undue importance to external appearance; in other words, to what has been termed *semeiotics*, which occupies a very considerable place in the medical treatises of the Asclepiadæ. Indeed, the writings on this subject constitute more than one-eighth part of the entire Hippocratic collection. To prognosis, also, Hippocrates ascribed very great importance, saying that "The best physician is the one who is able to establish a prognosis, penetrating and exposing first of all, at the bedside, the present, the past, and the future of his patients, and adding what they omit in their statements. He gains their confidence, and being convinced of his superiority of knowledge they do not hesitate to commit themselves entirely into his hands. He can treat, also, so much better their present condition in proportion as he shall be able from it to foresee the future," etc.

To the careful scrutiny of facial appearances, the position, and other body-marks about the patient he attributed very great importance; in fact, so positive was he about these matters that he embodied the principal rules of semeiotics into aphorisms, to which, however, there came later so many exceptions that they lost much of their value. From certain passages in his book on *Prediction*, and from the book on *Treatment*, which is a part of the Hippocratic collection, it appears that it was the custom then of physicians to announce the probable issue of the disease upon the first or second visit,—a custom which still prevails in China and in Turkey. It gave the medical man the dignity of an oracle when right, but left him in a very awkward position when wrong.

To Hippocrates we are indebted for the classification of sporadic, epidemic, and endemic forms, as well as for the division of disease into acute and chronic. Hippocrates wrote extensively on internal disease, including some particular forms of it, such as epilepsy, which was called the sacred disease; also fragments on diseases of girls, relating particularly to hysteria; also a book on the nature of woman, an extensive treatise on diseases of women, and a monograph on sterility. That Hippocrates was a remarkably close observer of disease as it appeared to him his books amply prove; in fact, they almost make one think that close observation is one of the lost arts, being only open to the objection that too much weight was attached to insignificant external appearances, speculation on which detracted from consideration of the serious feature of the case. His therapeutics, considering the crude information of the time, was a vast improvement on that which had preceded, and really entitled him to his title of "Great Physician."

Of external diseases and their surgical therapeutics he wrote fully: on *The Laboratory of the Surgeon*, dealing with dressings, bandaging, and operating; on *Fractures;* and on *Articulations and Dislocations;* showing much more anatomical knowledge than was possessed by his contemporaries. The *Mochlic* was an abridgment of former treatises; in *Wounds of the Head* he formulated the dictum concerning the possible danger of trifling wounds and the possible recovery from those most serious, so often ascribed to Sir Astley Cooper. Other monographs, also, he wrote, on *Diseases of the Eye*, on *Fistula*, and on *Hæmorrhoids*. He described only a small number of operations, however, and all the Hippocratic writings on surgery would make but a very incomplete treatise as compared with those that belong to the next historical epoch; all of which we have to ascribe—in the main—to prejudice against dissection and ignorance of anatomy.

From the earliest times physicians and writers occupied themselves largely with obstetrics, as was most natural. The Hippocratic collection includes monographs on *Generation;* the *Nature of the Infant;* the *Seventh Month of Pregnancy;* the *Eighth Month of Pregnancy;* on *Accouchement;* on *Superfœtation;* on *Dentition;* on *Diseases of Women;* on *Extraction of the Dead Fœtus.* The treatise on superfœtation concerned itself mainly with obstetrics.

On epidemics Hippocrates writes extensively, showing that he had studied them carefully. He was among the first to connect meteorological phenomena with those of disease during given seasons of the year, expressing the hope that by the study of storms it would be possible to foresee the advent of the latter, and prepare for them. Seven books of the Hippocratic collection bear the title of *Epidemics*, although only two of them are exclusively devoted to this subject. In these books were contained a long list of clinical observations relating to various diseases. They constituted really a clinical study of disease.

The collection of Hippocrates's *Aphorisms* fills seven of the books; no medical work of antiquity can compare with these. Physicians and philosophers of many centuries have professed for them the same veneration as the Pythagoreans manifested for their golden verses. They were considered the crowning glory of the collection. Even within a short time past the Faculty of Paris required aspirants for the medical degree to insert a certain number of these in their theses, and only the political revolution of France served to cause a discontinuance of this custom. These aphorisms formed, says Littré, "a succession of propositions in juxtaposition, but not united." It has always been and always will be disadvantageous for a work to be written in that style, since such aphorisms lose all their general significance; and that which seems isolated in itself becomes more so when introduced into modern science, with which it has but little practical relationship. But not so if the mind conceive of the ideas which prevailed when these aphorisms were written; in this light, when they seem most disjoined they are most related to a common doctrine by which they are united, and in this view they no longer appear as detached sentences.

The school of the Asclepiadæ has been responsible for certain theories which have been more or less prominent during the earlier historical days. One of these which prevailed throughout the Hippocratic works is that of *Coction* and *Crisis*. By the former term is meant thickening or elaboration of the humors in the body, which was supposed to be necessary for their elimination in some tangible form. Disease was regarded as an association of phenomena resulting from efforts made by the conservative principles of life to effect a coction,—*i.e.*, a combination of the morbific matter in the economy, it being held that the latter could not be properly expelled until thus united and prepared so as to form excrementitious material. This elaboration was supposed to be brought about by the vital principles, which some called nature (*Physis*), some spirit (*Psyche*), some breath (*Pneuma*), and some heat (*Thermon*). The gradual climax of morbid phenomena has, since the days of Hippocrates, been commonly known as *Crisis*; it was regarded as the announcement of the completion of the union by coction. The day on which it was accomplished was termed *critical*, as were also the signs which preceded or accompanied it, and for the crisis the physician anxiously watched. Coction having been effected and crisis occurring, it only remained to evacuate the morbific material—which nature sometimes spontaneously accomplished by the critical sweat, urination, or stools, or sometimes the physician had to come to her relief by the administration of diuretics, purgatives, etc. The term "critical period" was given to the number of days necessary for coction, which in its perfection was supposed to be four, the so-called quarternary, while the septenary was also held in high consideration. Combination of figures after the Pythagorean fashion produced many complicated periods, however, and so periods of 34, 40, and 60 days were common. This doctrine of crisis in disease left an impress upon the medical mind not yet fully eliminated. Celsus was the most illustrious of its adherents, but it can be recognized plainly in the teachings of Galen, Sydenham, Stahl, Van Swieten, and many others. In explanation, it must be said that there have always existed diseases of nearly constant periods, these being nearly all of the infectious form, and that the whole "critical" doctrine is founded upon the recognition of this natural phenomenon.

The Hippocratic books are full, also, of the four elements,—earth, water, air, and fire; four elementary qualities,—namely, heat, cold, dryness, and moisture; and the four cardinal humors,—blood, bile, atrabile, and phlegm.

Owing to the poverty of knowledge of physics and chemistry possessed by the ancients, and notwithstanding their errors and imperfections, the doctrine of Dogmatism, founded upon the theory of coction and humors, was the most intelligible and complete among the medical doctrines of antiquity, responding better, as it did, to the demands of the science of that day. That Hippocrates was a profound observer is shown in this: that he reminds both philosophers and physicians that the nature of man cannot be well known without the aid of medical observation, and that nothing should be affirmed concerning that nature until by our senses we have become certain of it. In this maxim he took position opposed to the Pythagorean doctrine, and included therein the germ of a new philosophy of which Plato misconceived, and of which Aristotle had a very faint glimpse.

Another prominent theory throughout the Hippocratic books is that of Fluxions, meaning thereby about what we would call congestions, or conditions which we would say were ordinarily caused by cold, though certain fluxions were supposed to be caused by heat, because the tissues thereby became rarefied, their pores enlarged, and their humor attenuated so that it flowed easily when compressed. The whole theory of fluxion was founded on the densest ignorance of tissues and the laws of physics, the body of man being sometimes likened to a sponge and sometimes to a sieve. The treatment recommended was almost as crazy as the theory. Certain other theories have complicated or disfigured the Hippocratic writings, and certain have been founded on the consideration of two elements—*i.e.*, fire and earth—or on the consideration of one single element which was supposed to be air,—the breath, or *pneuma;* and there was—lastly—the theory of any *excedent*, which is very vague; of all of these we may say that they are not of sufficient interest to demand expenditure of our time.

The eclat which the second (*i.e.*, the Great) Hippocrates gave to the school of Asclepiadæ in the Island of Cos long survived, and many members of his family followed in his footsteps. Among his most prominent successors were *Polybius, Diocles*, and *Praxagoras*, also of Cos,—the last of the Asclepiadæ mentioned in history. Praxagoras was distinguished principally for his anatomical knowledge; like Aristotle, he supposed that the veins originated from the heart, but did not confound these vessels with the arteries, as his predecessors had done, but supposed that they contained only air, or the vital spirit. It has been claimed that he dissected the human body. He laid the foundation of sphygmology, or study of the pulse, since Hippocratic writers rarely alluded to arterial pulsations and described them as of only secondary importance.

The predominating theory in the Island of Cos was that which made health dependent on the exact proportion and play of the elements of the body, and on perfect combination of the four cardinal humors. This was the prevailing doctrine,—*i.e.*, the Ancient Medical Dogmatism, so named because it embraced the most profound dogmas in medicine, and was taught exclusively until the foundation of the school at Alexandria.

Two men, however, more commonly ranked among philosophers than among physicians of antiquity, dissected the statements of Hippocrates, and embodied them more or less in their own teachings, and thus exercised a great influence on the progress of the human mind, particularly in the direction of medical study. The first of these was PLATO, profound moralist, eloquent writer, and most versatile thinker of his day or any other. He undertook the study of disease, not by observation (the empirical or experimental method), but by pure intuition. He seemed to have never discovered that his meditations were taken in the wrong direction, and that the method did not conduce to the discovery of abstract truths. He gave beauty an abstract existence, and affirmed that all things beautiful are beautiful because of the presence of beauty. This reminds one of that famous response in the school of the Middle Ages to a question: "Why does opium produce sleep?" the answer being: "Because it possesses the sleepy principle." Plato introduced into natural science a doctrine of final causes. He borrowed from Pythagoras the dogma of homogeneity of matter, and claimed that it had a triangular form.

ARISTOTLE, equally great thinker with Plato, but whose mental activity was manifest in other channels, was born in Stagyrus, in Macedonia. He was fascinated by the teachings of Plato, and attained such eminence as a student that King Philip of Macedon made him preceptor to his son Alexander, subsequently the Great, by whom he was later furnished with sufficient funds to form the first known museum in natural history,—a collection of rare objects of every sort, transmitted, many of them, by the royal hands of his former student from the remote depths of Asia. Aristotle, by long odds the greatest naturalist of antiquity, laid the first philosophic basis for empiricism. He admitted four elements—fire, air, earth, and water—and believed them susceptible of mutual transmutation. He studied the nature of the soul and that of the animal body; regarded heat and moisture as two conditions indispensable to life; described the brain with some accuracy, but without the least idea of its true function; said that the nerves proceeded from the heart; termed the aorta a nervous vein; and made various other mistakes which to us seem inexcusable. Nevertheless, he was rich in many merits, and no one of his age studied or searched more things than he, nor introduced so many new facts. Although he never dissected human bodies, he nevertheless corrected errors in anatomy held to by the Hippocratic school. He dissected a large number of animals of every species, and noted the varieties of size and shape of hearts of various animals and birds. In other words, he created a comparative anatomy and physiology, and the plan that he traced was so complete that two thousand years later the great French naturalist Cuvier followed it quite closely. If he be charged with having propagated a taste for scholastic subtleties, he also furnished an example of patient and attentive observation of Nature. His history of animals is a storehouse of knowledge, and his disciples cultivated with zeal anatomy, physiology, and natural history. His successor, Theophrastus, was the most eminent botanist of antiquity.

It will thus be seen that Plato and Aristotle were the eminent propagators of two antagonistic opinions. One supposed knowledge to be derived by mental intuition, and the other that all ideas are due to sensation. Both count among moderns some partisans of the greatest acumen: Descartes, Leibnitz, and Kant being followers of Plato, and Bacon, Locke, Hume, and Condillac, of Aristotle.

The excuse for stating these things, which apparently do not so closely concern the history of medicine, must be that of the learned interpreter of the doctrine of Cuvier, that "The first question in science is always a question of method."

Hippocrates formed a transition between a period of mythology and that of history. His doctrine was received by contemporaries and by posterity with a veneration akin to worship. No other man ever obtained homage so elevated, constant, and universal. A little later ignorance reigned in the school that he made celebrated. Methods and theories were propagated there under the shadow of his name which he would have disowned.

Medical science now changes its habitation as well as its aspect, and from the record of Hippocrates and his work we turn to the fourth period of the Age of Foundation,—namely, the *Anatomic*, which extends from the foundation of the Alexandrian library, 320 B.C., up to the death of Galen, about the year A.D. 200.

CHAPTER II.

AGE OF FOUNDATION (*continued*).—*Anatomic Period:* Influence of the Alexandrian Library. Herophilus and Erasistratus. Aretæus, † B.C. 170. Celsus, A.D. 1-65 (?). Galen, † 200.—*Empiricism:* Asclepiades, B.C. 100 (?).—*Methodism:* Themison, B.C. 50 (?).—*Eclecticism.*

AGE OF TRANSITION, A.D. 201-1400.—*Greek Period:* Oribasius, 326-403. Ætius, 502-575. Alexander of Tralles, 525-605. Paulus Ægineta, 625-690.

FOURTH, OR ANATOMIC, PERIOD.— As already seen, Alexander the Great and his successors collected the intellectual and natural riches of the universe, as they knew them, and placed them at the disposal of studious men to benefit humanity; their complete value has not yet been exhausted, and never can be. This undertaking was carried out under conditions that made it one of extreme difficulty. Manuscripts were then rare and most costly; but few copies of a given work were in existence, often only one, and these were held almost priceless. Under these circumstances the establishment of a public library and of a museum was an act of philanthropy and liberality simply beyond eulogy, and did more to immortalize the founder of the collection than all his victories and other achievements.

This appears to have also occurred to two of Alexander's lieutenants—one Eumenes, Governor of Pergamos, and the other, Ptolemy, Governor of Egypt. After the death of the conqueror his generals shook of all dependence upon the central government, and endeavored to centralize their own authority. But these two were the only ones among so many leaders who did not devote all their attention to armies and invasion, but interested themselves in commerce and arts. So active were they in the enterprise that Eumenes had gathered two hundred thousand volumes for the library at Pergamos, and Ptolemy six to seven hundred thousand for that of Alexandria. The latter was divided into two parts, the greater and the lesser, the latter of which was kept in the temple of Serapis, hence known as the *Serapium*. These notable efforts to found enormous collections first excited praiseworthy rivalry among contemporaries and rulers, which, however, degenerated into contemptible jealousy, so that some of the rulers of Alexandria even went so far as to interdict the exportation of papyrus, in order to prevent the making of copies for the library of Pergamos. But the effect was unexpected, since it led to the invention of the paper of Pergamos, otherwise called parchment, which completely displaced the bark from which papyri were made. Be this as it was, the collection at Alexandria had a much more marked influence on the medical study of the future than that of Pergamos, and calls for our particular notice. About it sprang up first a collection of learned men, and then the inevitable result—a school of learning. It was Ptolemy Soter who called around him the most renowned men of his day. He provided them with homes adjoining the library, endowed them with salaries, and charged them with the classification and collation of manuscripts, or with the giving of instruction by lectures and discussions. Ptolemy himself sometimes took part in these feasts of reason, which became still more frequent and formal under his son Ptolemy Philadelphus. These were called the Feasts of the Muses and of Apollo,—*i.e.*, *ludi musarum*,—and, consequently, the place where they were held came to be termed the "*museum*." Often the subjects for discussion were announced in advance, and those who gained the most applause received rewards in accordance with the merits of their work. Among those who enjoyed these advantages under the reign of these two Ptolemies are prominently named two physicians, Herophilus and Erasistratus, the latter said to be the grandson of Aristotle. It was under this Philadelphus that the Hebrew wise men translated into Greek the Holy Scriptures, which translation has since been called the Septuagint—so called because it is supposed to have been translated by the members of the Sanhedrim, which was composed of about seventy men, or because, according to another legend, it was translated by seventy-two men in seventy-two hours. These *savants* of ancient Egypt, thus supported by the dynasty of the Lagides, gave the first place to the science of medicine. As regards this study, the school of Alexandria eclipsed almost from its origin the ancient schools of Cos and Pergamos, and during its existence was the leading institution of its kind in the world. At the time of Galen it was sufficient to have studied there, and even to have resided a short time in Alexandria, to obtain the reputation of being a physician. Nearly all the scholars of these five centuries had received instruction in this school. The principal reason for its eminence in medical instruction was the practice of dissection of human bodies, which, under the Ptolemies, was allowed and recommended, and by which the science of medicine received an extraordinary impulse. Although the prejudice of Egyptians was very strong against those who touched a dead body, the Ptolemies themselves are said to have participated in this kind of anatomical study, thus destroying by their example the odium previously attached to dissection. Strange to say, however, the practice of dissection fell into disuse toward the end of this Anatomic Period, and scholars preferred to indulge in subtle metaphysical discussions rather than study human tissues. But the principal reason for giving up this practice was the Roman domination of Egypt, the Romans, inconsistently, being perfectly willing to see any amount of bloodshed in the arena, and all sorts of inhumanities practiced upon living human beings, but holding that contact with a corpse was profanation; so that not a single anatomist of reputation had his origin in ancient Rome. "If on any occasion," says Renouard, "a foreign physician attached to the king or general desired to avail himself of the occasions that were afforded to examine the structures of the internal parts of the human body, he was obliged to conceal and carry off during the night some body abandoned to the birds of prey." To complete the melancholy termination of the Anatomic Period, the labors of the writers of those days were all lost by the burning of the great library by Julius Caesar, which was the beginning of the chain of disasters with which Egypt was accursed under Roman dominion. Although Mark Antony, induced thereto by the endearments and solicitations of Cleopatra, transported the library of Pergamos to Alexandria, even this was unavailing to restore the position of the school, since the atrocious and imbecile Caracalla took from the pensioners of the museum their privileges of common residence and every other advantage, and suppressed all public exhibitions and discussions. I can mention but few of the names most eminent during this Anatomic Period, and but a short account of the life and work of each.

The first deserving of mention was Herophilus, who was born in Chalcedon about the end of the fourth century before Christ, and supposed to be the first to undertake systematic dissection of the human body. The so-called Torcular Herophili, or common meeting-place of the sinuses at the occiput, named after him, gives evidence of his influence upon the study of anatomy. He wrote on all departments of medical science, concerning the eyes, the pulse, midwifery, etc., as well as numerous commentaries upon the Hippocratic writings,—describing the membranes of the brain and its vessels, the choroid plexus, the ventricles of the brain, the tunics of the eye, the intestinal canal, and certain portions of the vascular system. He alluded to the thoracic duct without knowing its purpose, and gave a more accurate description of the genitalia than any previous writer. Strange to say, but little is known of his later life, and of his death absolutely nothing.

Erasistratus was the son of Cleombrotus, a student of Metrodorus, and lived for some time at the court of Seleucus Nicator, whose son, Antiochus, he healed of a secret ailment, which happened to be a desperate love-affair with his mother-in-law, Stratonice. He wrote extensively on fevers, hygiene, paralyses, therapeutics, and many other subjects; regarded most diseases as due to overindulgence in food, which is not digested, and consequently putrefies. Plethora was for him the prevailing disease, against which he employed not only venesection, but fasting, and bandaging of the extremities. He was a diligent student of anatomy, and carefully described the brain in many of its grosser features, regarding it as the seat of the soul and the centre of the nerves. He also described more exactly than his predecessors the valves of the heart, which organ he regarded as the origin of veins and arteries. He discovered the lymph-vessels, and maintained, against Plato and others, that the epiglottis prevents the entrance of fluids into the lungs, but he supposed digestion to be produced by mechanical trituration in the stomach, and preferred gymnastics, exercise, diet, and baths to drugs or other therapeutic measures. He died about 280 B.C.

Aretaeus, who died about 170 B.C., was one of the most brilliant lights of antiquity previous to the Christian era, but, in spite of all this, of his life very little is known. He came from Cappadocia about the end of the reign of Nero, and lived in Alexandria. That he lived in Alexandria is apparent from his numerous references to its location, to the habits and therapeutics of the Egyptians, and to the geography of the country. Furthermore, references to its diseases abound in his writings, so that it is made to appear that he had had the best advantages there, although he must have traveled extensively. But a small portion of his writings remain, and these consist, for the most part, of compendiums of pathology and therapeutics. He described disease, not in anatomical order from head to foot, but under the classification of acute and chronic. With the exception of Hippocrates, he has shown himself the most free from vague, arbitrary speculation, and from the dogmatism of the schools of any writer of antiquity. He, more than any other up to his time, endeavored to found pathology upon a sound anatomical basis. For every picture of disease he endeavored to provide a suitable anatomical accompaniment. This appears particularly, for

FIG. 4.—AULUS CORNELIUS CELSUS.

instance, in his description of intestinal ulcers due to dysentery, or the paralyses following brain affections, or his description of pharyngeal diphtherias, of which he gave a good account under the name of Syriac or Egyptian ulcers. Pulmonary tuberculosis, tetanus, and anal fistula are amply mentioned in his writings. His therapeutics were simple and rational; he laid great stress upon dietetic treatment. His surgical writings appear to have all been lost, but there is every reason to think that he brought to bear upon external medicine the same good sense which he applied to internal affections.

Cornelius Celsus, the most celebrated author for a number of centuries, was born in Rome about the time of Christ. Brilliant as he was, he exerted a wide-spread influence for centuries. The exact date of his death is unknown. He was a contemporary of the greatest philosophers, poets, and *savants* of Rome during its most brilliant period. He studied rhetoric, philosophy, the art of war, economics, and medicine—he was, in fact, a walking encyclopædia of the knowledge of his day; but it is in medicine that he shows to best advantage, and in his capacity as a physician he was and is best known. The direction in which Celsus appears to least advantage is in failure of power of direct observation, and in yielding unquestioning obedience to the views and dicta of Hippocrates, for whom he possessed the greatest reverence, not being able to brook any serious contradiction or opposition to his opinions. In this reverence for Hippocratic authority he was followed by many less prominent successors, the consequence being a failure to train men as observers, the endeavor being to make them simply storehouses of information derived from Hippocratic writings. As a result, Celsus wrote but little, or else his writings are lost. He contented himself mostly with a mere commentary upon the writings which he so highly revered. But little of his writings remain, and these pertain mostly to the therapeutics of curable disease, dietetic, pharmaceutical, and surgical. Although he exercised great authority during his period, he was later totally supplanted by Galen, and his views are seldom mentioned in the writings of those subsequent to this great physician. His death must have taken place during the first century after Christ.

Of all the students of Hippocratic dogmatism, the most earnest, skillful, and learned was Claudius Galen, a native of Pergamos, a place already celebrated for its temple dedicated to Æsculapius, for its school of medicine, and for a library which had been removed to Alexandria. He was placed by his father under the most distinguished teachers in all of the sciences, and even as a young man showed extraordinary progress, and became early a disputant with the most erudite in grammar, history, mathematics, and philosophy. He has related how in two different dreams

FIG. 5.—THE CONVERSION OF GALEN.
(From an old engraving published by R. Sayer & J. Bennett, London, 1775.)

he was urged by Apollo to study medicine. He traveled widely for instruction, and remained some time in Alexandria. On his return to his own country he was charged by its ruler to dress the wounded in the great circus, which furnished him opportunity for displaying all his anatomical knowledge and surgical skill. Not remaining long at home, he went to Rome, where his renown had preceded him, and where, by his brilliant elocution, his accurate logic, and his profound erudition, as well as his versatility and practical skill, he at once took the highest place. But here his rapid success, his vanity, his disdain for his colleagues, and his useless boasting, as well as his natural jealousy, gained him the enmity of nearly all his contemporaries, and his stay at Rome was thereby made very disagreeable. In his work on *Prenotions* he accuses his colleagues of base jealousy and stupid ignorance, lavishes upon them such epithets as "thieves" and "poisoners," and closes by saying that after having unmasked them he would leave them to their evil designs by abandoning the great city to seek a home in a smaller place, where the surroundings would be to him more congenial. This threat he carried out, but soon returned to Rome upon the invitation of the Emperors Marcus Aurelius and Lucius Verus, whose confidence, as well as that of their successors, he enjoyed. He is supposed to have lived to the age of seventy-one, and to have died about 200 A.D. Galen strongly denied being attached to any of the sects of his day, and regarded as slaves those who took the title of Hippocratists, Praxagoreans, Herophilists, and so on. Nevertheless, his predilection in favor of the Hippocratic writings is well marked, for he explains, comments upon, and expands them at length, refutes the objections of their adversaries, and gives them the highest place. He says: "No one before me has given the true method of treating disease; Hippocrates, I confess, has heretofore shown the path, but as he was the first to enter it he was not able to go as far as he wished. . . . He has not made all the necessary distinctions, and is often obscure, as is usually the case with ancients when they attempt to be concise. He says very little of complicated diseases; in a word, he has only sketched what another was to complete; he has opened the path, but has left it for a successor to enlarge and make it plain." This implies how he regarded himself as the successor of Hippocrates, and how little weight he attached to the labors of others. He held that there were three sorts of principles in man: spirits, humors, and solids. Throughout his metaphysical speculations Galen reproduces and amplifies the Hippocratic dogmatism. Between perfect health and disease there were, he thought, eight kinds of temperaments or imperfect mixtures compatible with the exercise of the functions of life. With Plato and Aristotle, he thought the human soul to be composed of three faculties or parts: the vegetative, residing in the liver; the irascible, having its seat in the heart; and the rational, which resides in the brain. He divided diseases of the solids of the body into what he called *distempers;* he distinguished between the continued and intermittent fevers, regarding the quotidian as being caused by phlegm, the tertian as due to yellow bile, and the quartan as due to atrabile. In the doctrine of coction, crises, and critical days he agreed with Hippocrates; with him he also agreed in the positive statement that diseases are cured by their contraries. From all this it will be seen that Galen must be regarded as one of the earliest of Hippocratic dogmatists. He was a most extensive writer, and it is said that the total number of his works exceeded one hundred. His contributions to anatomy were not insignificant. For myology he did a great deal. He wrote a monograph on the skeleton in which he recommended that bones be seen and handled, not merely studied from books, and that the student should go to Alexandria, where teachers would place before him the real human skeleton. It has been inferred that there was not, in his time, in Rome a single skeleton. He wrote fifteen books on anatomy, of which six are lacking; also an extensive treatise on the lesions of the human body, distributed among seventeen books which have come down to us. He is supposed to have introduced the term "symphysis," and he described nearly every bone in the human body. By him the muscles were no longer considered as inert masses and tissue-layers serving to cover the bones, but he classified them according to their distinct functions, and studied separately their form and uses. The location of the vessels and nerves between them was also noted, and it was proved that muscles were indispensable to the accomplishment of voluntary motions. Galen was, perhaps, the first vivisector of all, since he exposed muscles of living animals, and showed how alternate tension and relaxation of distinct groups set the bones in motion, after the manner of levers; he named a great number of them, but, curiously, took no note of others. His classification according to their uses is followed down to the present day—*i.e.*, flexors, extensors, etc.

The Hippocratic authors confounded the arteries with the veins. Praxagoras first distinguished two kinds of vessels which he supposed to contain air, whence the name artery. Aristotle and Erasistratus maintained this view, which prevailed until the time of Galen, who devoted a book to the refutation of it, basing his argument upon the observation that always when an artery is wounded blood gushes out. How near he came to being the discoverer of the circulation may thus be seen. A little less reverence for authority and a little more capacity for observation would have placed him in possession of the knowledge, lack of which for so many centuries retarded the whole profession. He thought the veins originated from the liver—in this respect being behind Aristotle—but considered the heart as the common source of the arteries and veins. Even the portal system of veins confused him, and he erroneously described a superior and inferior aorta, but atoned for this by describing the umbilical veins and arteries. Aristotle also had supposed all the nerves originated from the heart, but Galen stated that they are derived from the brain and spinal marrow, and pointed out two kinds of nerves: those of sensation, which he thought proceeded from the brain, and those of motion, which he considered to originate in the spinal marrow. Thus, he described distinct nerves of sensation and motion, but sadly confused their anatomy. He seems also to have had some notion of the great sympathetic, although it was by no means accurate. He suggested the division of the principal nerves, in order to prove the fact that nervous energy is transmitted from the encephalon to other parts of the body. He speaks of glands, and thought they discharged their secretions through veins into the various cavities, but regarded them rather as receptacles of excrementitious matter than as agents for secretion of valuable fluids. He even regarded the mammæ as glandular bodies in this sense, although he knew, of course, the value of their secretion. To Galen we owe the division of the body into cranial, thoracic, and abdominal cavities, whose proper viscera and envelopes he described. He spoke of the heart as having the appearance of a muscle, but differing from it. He regarded it as the source of natural heat, and the seat of anger and of violent passions. He appreciated that inspiration is carried on by enlargement of the thoracic cavity. He thought that atmospheric air entered the cavity of the cranium through the cribriform plate of the ethmoid and passed out by the same route, carrying with it excrementitious humors from the brain, which were discharged into the nasal fossæ. But some portion of air thus entering remained, according to his views, and combined with the vital spirits in the anterior ventricles of the brain, from which combination originated the animal spirits and immediate agents of the rational soul. These acquired their last attenuation in the fourth ventricle, whence they would pass out drop by drop through a round, narrow tube.

From this brief *résumé* of the anatomy and physiology of Galen it will be seen that by the end of the second century of the Christian era immense progress had been made since the foundation of the Alexandrian school, and that it was due to the impetus in the study of anatomy given by

Herophilus and Erasistratus, who not only made numerous dissections, but resorted to frequent vivisections. It is even said that Herophilus did not hesitate to employ his knife on live criminals who were subjected to him for experiment; but this has been a popular tradition about almost every anatomist of antiquity, and there is no evidence in confirmation of the unkind rumor, although that such experiments might be legally and justly performed has occurred to the minds of many. But zeal for dissection rapidly cooled off, and Galen barely mentions five or six men who devoted themselves to it in the space of nearly four hundred years down to his time. He speaks of *Rufus* of Ephesus,—who lived under Trajan of *Marinus*,—who wrote in the beginning of the second century A.D., and of *Quintus*, who instructed his own preceptor. None of them left a reputation, however, approaching that of Herophilus and Erasistratus, with whom Galen alone could compare by the number of his experiments and his discoveries. Galen strove as hard as one of his position might, by example and precept, to awaken in his contemporaries a desire for anatomical knowledge, but could not overcome their indifference. After him the practice of dissection appears to have been lost, either from the redoubled prejudices of the superstitious, who opposed it, or as the result of the apathetic ignorance or the ignorant apathy of the physicians.

It has been shown that, during the Hippocratic era and subsequently, the physicians even of primitive times followed more or less by instinct the empirical method. Acron of Agrigentum was a contemporary of Pythagoras, and affirmed that experience is the only true foundation of the healing art. Hippocrates, however, showed himself more anxious to report faithfully clinical facts than to dispute theoretical views.

The surprising progress in anatomy and physiology made during the first portion of the Anatomic Period and during the better days of the Alexandrian institute did not keep men from confounding several different points in the Hippocratic doctrine, by which confidence in the same was naturally shaken. Thus many new speculations were hazarded which nullified each other. In the midst of this confusion practitioners continued to seek in experience a refuge from the incessant variations of dogmatism and the sterile incertitude of the skeptics. Thus, empiricism as a school of practice became placed upon a firmer and firmer foundation, and the empirics of that day seem to have laid the true basis of our art. Their doctrine took at first a rapid growth, and Galen spoke of it with great regard. The circumstances under which it was proclaimed were most favorable for its propagation. Theories had fallen into confusion; practice, methods, and opinions were questionable. Everything was conjecture, and that which rested on the evidence of facts was by the empirics received with enthusiasm. Although founded on pure observation, it did not put an end to differences of opinion, and in the eyes of the ancients it lacked in solidity, because it did not attach itself to any philosophic theory then known. This doctrine was then best able to captivate physicians on account of its simplicity, contrasted with the general inability to satisfy speculative minds; but for this very reason it subsequently fell into disgrace, and the term "empiricism" became synonymous with ignorance. For centuries condemned and despised, it was revived from its long humiliation under the name of the Experimental Method, and achieved, after the labors of Bacon, Locke, and Condillac, almost universal dominion in the sciences.

This doctrine had been proclaimed for about a century during the period of which we now speak, but later led men into a fondness for secondary generalities or for the elevation and magnifying of trifles, which confused their minds and terminated its usefulness to science. Meanwhile, a man of great intelligence, renowned as an elocutionist, well versed in the doctrine of philosophers and grammarians—namely, ASCLEPIADES, of Bythinia—came to Rome with the intention of teaching rhetoric. By his talent and personal address he soon became one of the most illustrious persons in the Roman Republic; so early as 150 B.C. he enjoyed a high reputation as a rhetorician, and was one of the intimate friends of Cicero; nevertheless, he abandoned letters, undertook the practice of medicine, and sought moreover to create a new system, being unwilling to follow in the track of his predecessors. Imbued with the philosophy of Epicurus, who was then in high repute, he deduced from it a theory which was in harmony with the philosophy of the day. He thought that the elements of the body existed from eternity; that they were indivisible, impalpable, and perceptible to the reason only. These elements he named *atoms*, which were supposed to be animated by perpetual motion, and from which, by their frequent encounters and fortuitous contention, all sensible phenomena were supposed to result. He explained the properties of the body by saying that compounds were aggregates of atoms, differing very much from atoms themselves. Solid silver, he said, is white, but, reduced to powder, appears black; the horn of the goat, on the contrary, is black, but if it be razed its particles are white. This, it will be seen, was the parent of our present atomic theory. He ridiculed the theories of Hippocrates concerning coction, crises, etc., and sarcastically called the Hippocratic treatise on therapeutics "a meditation on death."

Asclepiades based his own therapeutics on endeavors so to enlarge the pores of the human body that disease could find egress, or so to constrict them that it could not enter; consequently he rejected all violent remedies, such as vomits, purges, etc., and his favorite remedies were hygienic,—for the most part bodily exercise.

A celebrated disciple of Asclepiades was THEMISON, of Laodicea (B.C. 50), who was led by the teachings of his master to lay the foundation of the so-called *Methodism* as opposed to *Dogmatism* in the school of Cos. By him and his followers a very arbitrary arrangement of diseases was made, according to what they considered the constrictive, or contractive; the fluxionary,—congested or relaxed; and the mixed forms. From this division of diseases it appears that, according to the methodists, there were only two kinds of therapeutic indications to follow,—namely, to relax where there was constriction, to constrict where there was relaxation. They, however, admitted a third creditable result, which they called prophylactic; but the pure methodists, such as Cœlius Aurelianus, admitted neither specific disease nor specific remedies, and erased from their materia medica purgatives, diuretics, emmenagogues, nauseants, etc.

According to the methodist doctrine, the study of medicine was so abridged that one of its prominent exponents said that he felt able to teach the whole of medical science in six months. It made rapid progress, and consequently was most attractive to the numerous young neophytes who were anxious to finish their apprenticeship and hasten into practice. It is not one of the smallest of the services which Galen rendered to his time and to posterity that he demolished the sophistry of the methodists, demonstrated the insufficiency of their practice, and brought to bear upon them the wittiest satire, calling them the asses of Thessaly, alluding thereby to their lack of literature and medical instruction.

In summing up, then, the basis for the various systems of medicine during this period of antiquity, it is seen that the most ancient doctrine of all—*Dogmatism*—directs our attention especially to the animal economy in health and disease; that it took account of the union of vital forces, of sympathies in the organism, and of nature's efforts to repel both internal and external deleterious influences, which providential tendency manifests itself especially in certain acute diseases. This was the strong side of dogmatism. Its weak side consisted in this: that it was held that the causes of diseases inhere in the access of certain qualities and humors along with organic forces,—such as dryness or moisture in combination with bile or atrabile,—and the treatment was directed against these supposed causes. It was on account of this weakness that the enemies of dogmatism attacked it. The empirics opposed the idea that inaccessible and occult causes of disease could become the basis for rational treatment. They affirmed that there was no consistent relation of antagonism or similitude between the disease and the remedies which cured it.

The *Methodists* somewhat improved on the doctrine of empiricism, but ran wild in its improvement and erected over their fundamental theory such a superstructure of secondary and tertiary generalities as to cause the fundamental part to be entirely obscured from sight.

There were not lacking, in those days of old, certain educated physicians who more or less vaguely comprehended that the entire truth of medicine did not inhere in any one of these systems, but that there was good and evil in each. These men, not being able to establish general rules, tried to decide practical questions according to their fancy or their reason. They assumed the name of *Eclectics* or *Episynthetics*, meaning thereby that they adopted no exclusive system, but selected from each that which seemed to them best. They did not constitute a sect, because they had no precise dogmas nor theories, but they should not be confounded with the Pyrrhonians, who held to doubt as a fundament doctrine, the true eclectic doubting only that which he could not understand. True eclecticism in medicine, however, is rather the absence of fixed principles, or, as Renouard says, it is "individualism erected into a dogma, which escapes refutation because it is deficient in principle." Many became eclectics to avoid discussing principles, and made of it a shelter. In one sense, then, an eclectic is one destitute of profound convictions, who sides with no particular party, is committed to no person or doctrine, and who is often so indifferent that he cannot judge with impartiality; consequently, to be truly eclectic is different from being an adherent of a school of eclecticism.

During the historic period just reviewed, anatomy and physiology made most progress, next internal and external nosography, and next to these medical and surgical therapeutics, and although Cœlius Aurelianus and Aretæus have left to us by far the best books issued up to their times, nevertheless not one of the writers of this period has achieved the distinction in which Hippocrates is held, since he, perhaps more than any other, combined intelligence, sincerity, disinterestedness, love of his art, and humanity.

Under the classification of Renouard, already alluded to, the so-called *Age of Transition* includes centuries commencing with the death of Galen, about A.D. 201, and ending with the revival of letters in Europe, about the year 1400. The first period of this transition age is the so-called Greek Period, which ends with the burning of the Alexandrian library, A.D. 640.

At the time when this historic period commenced all the known world was under the dominance of a single man. The power of Septimus Severus had more extent than that of Alexander the Great, and bid fair to be of a much longer existence. The Roman dominion, cemented by seven hundred years of bold and persevering government, seemed almost immovable. While the savages upon its frontiers occasionally troubled its peace, none were strong enough to penetrate its centres or place it in real peril. The great civil wars had ceased, or changed their object.

Both the people and the senate, those two eternal competitors, had gotten over the struggle for supreme power; monarchial government was accepted as a matter of fact, and the citizens contended only for choice of a master.

Similar changes had taken place in the domain of the mind; philosophical discussions, which were so essentially a part of the schools of the ancient Greeks, had nearly lost their interest and were being discontinued. Such disputes as took place related less to principle than to interpretation of the language of the teacher. In morals, Plato, Epicurus, and Zeno were followed until the principles of Christianity gradually supplanted their teaching; in physics and metaphysics the authority of Aristotle, and in medicine that of Galen, were simply undisputed.

Conditions being such as these, there was naturally but one sect in medicine, and one method of study and practice. Medical science retrograded rather than progressed, sad to say, and was undisturbed by any remarkable revolution. The scepter of medicine passed from the hands of one nation to those of another, and the language of Hippocrates and Galen was later replaced, as will duly be seen, by that of Avicenna and Albucassis. But this Greek Period, which is one of transition, offers little for our consideration more than the lives and writings of four of its most eminent physicians, who by their study in the school of Alexandria, and by their writings and teachings, left reputations which were sustained until the invasion of the Arabs. Of these it may be said that, while they did little or nothing original, and simply commented upon the writings of Hippocrates and Galen, they kept burning the torch of medical learning which else had been almost extinguished by their indolent contemporaries. Of these various commentators—for they were little more than that—the first of any importance after Galen was *Oribasius*, who was born in Pergamos (326–403); he early attached himself to the fortunes of Julian the Apostate, and followed him into Gaul when

he was made its governor. Julian appreciated the good
qualities of Oribasius, made him an intimate friend, and
after he himself became emperor appointed his friend
as quæstor at Constantinople. After the emperor's untimely death, Oribasius remained faithful to his memory,
but his jealous colleagues so falsely and so successfully
misrepresented his fidelity that he was disgraced, spoiled of
his office and property, and banished among a barbarous
people. In this new field, however, he displayed such
courage, effected such extraordinary cures, discoursed so
eloquently, and so attached to himself the savage men
around him, that he was by them regarded as a god. The
fame of this homage in time reached the ears of the Emperors Valens and Valentinianus, who recalled him, reim-
bursed him for his losses, and permitted him to enjoy his
high reputation and fortune to the end of his days. He
was held to be the wisest man of his time, most skillful in
medicine, and the most charming in conversation. He
dedicated a collection of seventy books to Julian, his first
patron, and edited, at a later period, an abridgment of this
work for the benefit of his son. His principal merit consisted in reproducing the ideas of others with such clearness, order, and precision that the summaries that he gives
of them are often preferable to the originals. What he has
said of pregnant women, nursing, and the earliest education of the child has been copied literally by writers for
twelve centuries since his time. It must be said of him,
however, that his prepossession in favor of Galen was so
great that he adopted servilely his ideas and even his
words to such an extent that he has been surnamed "the
ape of Galen."

Ætius was born in Mesopotamia in the year 502 and
died in 575. He studied at Alexandria, and afterward went
to Constantinople, where he became a chamberlain at court.
Ætius was the first medical man of any note who professed
Christianity, as is shown by such passages as this one: he

said that in the composition of certain medicaments the following words should be repeated in a low voice: "May the God of Abraham, the God of Isaac, and the God of Jacob deign to bestow upon this medicament such and such virtues." In another place he recommends that to extract a bone from the throat the following words be pronounced: "Bone—as Christ caused Lazarus to come forth from the sepulchre, as Jonah came out of the whale's belly—come out of the throat or go down." But he exhibits the same credulity in not doubting the miraculous virtues attributed by the quacks of his day to most remedies.

Like Oribasius, he collected everything that he found remarkable in the writings of his predecessors, and has preserved certain fragments of antiquity which would otherwise have been lost. His work formed a complete manual of medicine and surgery, except that it lacked anatomical descriptions and references to dislocations and fractures.

Alexander of Tralles (525–605), a city of Lydia, where Greek was spoken, was a son of the physician Stephen, and the most celebrated of five sons, who were all distinguished for their learning. He traveled extensively, and fixed his residence in Rome, where he became celebrated. He lived to an advanced age, and, being no longer able to practice, composed a treatise of twelve books, exclusively devoted to affections that did not require the aid of surgery. He professed the greatest veneration for Galen, but did not blindly adopt his opinions. He described the first reported case of excessive hunger and pain due to intestinal worms; he advised venesection in the foot rather than in the arm; but with all his sound judgment and mental enlightenment he had faith in amulets and talismans, and widely recommended them. It may be said for him, such was the universal prejudice of his age, the whole world being plunged in superstition, that it was necessary for every one to pay some tribute to the prevailing belief; and we may add that it is necessary to make this excuse for some who practice much nearer to ourselves than did those ancient physicians.

Paul, or *Paulus*, surnamed *Ægineta* (because he was born in the Island of Ægina), was among the last of the Greek physicians who have special interest for us. It is supposed that he died about A.D. 690. He traveled extensively, and his skill in surgery and obstetrics rendered him celebrated even among the Arabs, whose midwives sent for him in consultation from great distances. He composed a compendium of medicine, divided into seven books, and not only did not hesitate to borrow from his predecessors, but quoted from them most extensively; a number of his chapters were taken almost *verbatim* from Oribasius; however, he made no secret of it, but rather boasted that he had judiciously sought to appropriate the best of the writings of those he most revered. He showed originality, however, in the treatment of hydrocephalus, in advising paracentesis of the thorax and abdomen, in the extraction of calculi from the bladder, in the treatment of aneurism, the excision of hypertrophied mammæ in men, etc. He was the first to describe varicose aneurism, and the first to perform the operation of bronchotomy after the method borrowed from *Antyllus*, of which he has transmitted a very detailed account. Of this Antyllus, by the way, it may be added, *en passant*, that he was one of the most distinguished and original surgeons of antiquity. He flourished during the third century after Christ; was the first to describe the extraction of small cataracts; and is, perhaps, best known to the surgical world to-day by his exceedingly bold plan of opening aneurisms, so successfully imitated a generation or so ago by James Syme.

It has already been seen that before and during the early centuries of the Christian era the secrets and learning of the physicians tended to pass gradually into the hands of the priests. It was so in the temples of ancient Greece, it was so in Alexandria, it became so in Rome, it has been so even in modern times, although only for brief periods of time. This has come about in some measure from the cupidity of the clerical orders, partly because it required a certain amount of intelligence and knowledge to become a priest, and partly because, owing to ignorance, credulity, and superstition, diseases have at all times been regarded by the ignorant as evidence of divine wrath and chastisement, or of diabolical or occult influences, rather than the effect of natural causes. Hence men have turned ever toward prayers, exorcism, and expiation, especially when exhorted thereto by the priests. This has been the sacerdotal aspect of the practice of medicine in all times, and when the priests have usurped therapeutic functions they have done harm rather than good. So long as theology and science work hand in hand, each redounds to the credit of the other, but always in the history of man when theology has appropriated that which did not belong to it it has brought ridicule upon itself and has delayed the progress of knowledge. There have been frequent rebellions against religious authority in ancient as in modern times. For instance, at the commencement of the fifth century before Christ the Pythagoreans were dispersed, and the doctrines of Cos and Cnidus—*i.e.*, the Hippocratic teachings—were promulgated; and, again, in the course of events, when the descendants of Æsculapius became servile attendants at the temple and adjuncts to the priesthood or a part of it. At first, in Alexandria, the physicians were supreme; their disciples, however, had the same blind reverence for authority that too many workers in the field of theology have evinced, and men once more practiced medicine on the traditions of the past, and in so doing allied themselves more and more to the temples.

In Rome, at first, the oldest and best instructed of the relatives treated the diseases of his family as he understood them; simply shared this duty with its other members. Cato, the censor, was much engrossed with this domestic medicine; he wrote a book in which he recommended cabbage as a sovereign remedy in many diseases. He venerated the number 3, as did the Pythagoreans; did not disdain to transmit to posterity certain medical words which it was believed should be repeated to assist in the reduction of dislocations and fractures. This old censor seemed to have a profound hatred for medical men, and most absurd ideas of their works and claims, although doubtless many Greek physicians who came to Rome merited the invectives which he launched against them. Then came Asclepiades, of Bythinia, as already mentioned, whose talents were far superior to those of his Roman contemporaries, and who did not need to call to his aid charlatanism and deceit. This medical hero unfortunately had many worthless and dishonest imitators, who appealed to superstition and ignorance in every dishonest way, and who desired to be judged by the luxury and elegance they displayed. Hence for a long time in Rome medicine was practiced without license. The Emperor Anthony the Pious was the first to occupy himself with regulating the practice of medicine. He granted certain immunities, but did ask for proof of qualifications. A certain physician to Nero, Adromachus, was honored by the emperor with the title of Archiater,—*i.e.*, royal healer,—but Galen, who was physician to Marcus Aurelius, never bore it. From the time of Constantine the Great, however, the title is frequently met with in the edicts of the emperors. In fact, there were two sorts of these,—one named the Palatine, who belonged to the household of the reigning monarch and who held high rank among the nobility; and the other called the Popular Archiaters, who were public-health officers. No one could practice medicine in the jurisdiction of one of these without examination and authorization. Those who transgressed this regulation were punished with a fine of two thousand drachmas. The Popular Archiaters were pensioned by the city, enjoyed certain privileges, and had to attend the poor gratuitously. Practitioners who were not members of the College of Archiaters had no pay, no rights, nor emoluments. The Popular Archiaters were elected by the citizens from many candidates who had proved their capacity before the college of this medical organization. The evils of medical anarchy were thus remedied; this happy condition existed until the empire was broken up by barbarism.

It is during this period—about 400 A.D.—that we first find a class of citizens to whom was delegated the duty of preparing drugs ordered by physicians. Their duties were in some respects similar to those of our apothecaries, although in attainment and in social position they were far below the physicians. They were termed *pharmacopolists*.

It is worth while to stop a moment to inquire what were the medical charitable institutions of antiquity. Even in the days of ancient Athens there was a certain gymnasium, called the Cynosarga, in which abandoned and illegitimate children were brought up at public expense until such time as they were able to serve their country. A little later several private institutions of this kind were established. Rome in her earlier day never had such institutions. To be sure, she distributed provisions, or else remitted taxes, to parents who were unable to support their children, or even permitted them to destroy their newborn children when unable to maintain them; but there were no bonds of sympathy which induced the patricians to succor the plebeians in time of disease and distress; slaves were cared for as were cattle. It is one of the debts we owe Christianity that, under its influence, the first almshouses and retreats were established in Rome. It has been said that the Emperor Marcus Aurelius first instituted anything like a dispensary service in the Sacred City. We are told, also, of an illustrious woman, St. Pauline, living in the midst of the greatest wealth and pomp, who retired from society and devoted her life to charity and self-denial. She went to Jerusalem, united with other Christian women of the same mission, and formed, under the direction of St. Jerome, a sisterhood whose members divided their time between reading sacred books and doing good works. They offered an asylum for the faithful and a hospice for the benefit of the indigent sick, and even established a home for convalescents outside the city-walls. After the model thus set, heathen emperors, Christian kings, and Moslem caliphs showed their zeal in this good direction by the erection of sumptuous edifices and other rich endowments for the relief of suffering human beings.

Reviewing now the Greek period, let it be remembered that in the time of Galen animals were dissected, and that he made anatomical demonstrations on monkeys; that sometimes the corpses of the enemy were rudely dissected upon the field of battle, but that finally the practice of dissection fell into disuse, and human anatomy was studied only from books, the early Christians having evinced even more horror of the dead body for the purposes of anatomical study than did their pagan predecessors, while the Fathers of primitive times launched their anathemas against the dissection of human remains. Here, again, as usual, the interference of the church worked only general harm. This abandonment of anatomy contributed doubtless to the decadence of medicine; by the rapid extension of Christianity the pagan schools were disorganized and broken up, the profane sciences (such as medicine) were discarded, and the teachers still remaining in the old schools were ruined. Passion for religious controversy was engendered and took the place of study or original research, even to such an extent as to hasten the fall of the Empire of the East. In addition to these factors, reverence for authority of the past—that terribly oppressive weight which has kept down so much which would otherwise have risen early, and which has been the greatest enemy of human learning—permitted the explanation of natural phenomena to be sought only in the writings of revered ancients, and not in living beings. No one dared to advocate changes in regard to received doctrines, and there could be no such thing as progress. Only two men in the lapse of four centuries showed any originality; these were Alexander of Tralles and Paul of Ægina, whose lives have already been briefly rehearsed. It is with some relief, however, that we can think that this period, so unfruitful in scientific progress, was not so in social amelioration. By the organization of the institutions above alluded to charlatanism was checked, by the requirement of capability and good character society was benefited, and the charitable institutes of this epoch perhaps gave the world its best models in teaching and an insight into the most valuable means of medical instruction. Of the old Greek Period, then, we may say that it accrues rather to the benefit of humanity than to that of science.

CHAPTER III.

AGE OF TRANSITION (*continued*).—*Arabic Period:* A.D. 640-1400. Alkindus, † 873. Mesue, 777-857. Rhazes, 850-932. Haly-Abas, † 994. Avicenna, 980-1037. Albucassis, † 1122. Avenzoar, 1113-1161. Averroës, 1166-1198. Maimonides, 1135-1204. *School of Salernum:* Constantinus Africanus, 1018-1085. Roger of Salerno, 1210. Roland of Parma, 1250. The Four Masters, 1270 (?). John of Procida.

The Arabic Period, which began with the second destruction of the Alexandrian Library—640 A.D.—ends with the fourteenth century. At the commencement of this period the Roman Empire of the West scarcely existed: the magnificent territory which composed it had been overrun and subdued by barbarous tribes from the forests of the North, while from its ruins had risen several independent kingdoms,—that of the Franks in Gallia, of the Visigoths in Spain, and of the Lombards in Italy. The last of the Western emperors of note was Justinian, whose army and generals—especially the genius and heroic devotion of Belisarius—threw some glory upon Italy, Sicily, Africa, and Spain. Meantime the Empire of the East, surrounded by enemies, and harassed from all directions, still sustained itself with vigor. The Turks had begun to show themselves on the banks of the Danube; those eternal enemies of Rome—the Persians—made incessant war; and a new and terrible enemy had sprung up in the deserts of Arabia. Then came one who was at the same time legislator, prophet, and conqueror, and united under one faith and one leader tribes hitherto divided and warring against each other. Thus arose a powerful and enthusiastic nation, animated by thirst for conquest and ardor for proselytism. In less than a century after the first preaching of Mahomet, all of Arabia, India, Syria, and Egypt were in the hands of his followers. In the year 640 Amrou effected the conquest of Egypt, seized Alexandria, and the great library of five hundred thousand volumes was, by order of Omar (successor to Mahomet), delivered over to the flames; and the historian Abulpharagius declares that these books served for six months to heat the public baths, four thousand in number. Such were the first fruits of the establishment of Islam.* Happily, zeal of proselytism somewhat abated among the Mussulman princes, and religious fervor gave place to policy; so that the later Arabian caliphs showed themselves, in general, the protectors of the arts and sciences. Some, indeed, endeavored to collect the *débris* of the scattered treasures that had been so fortunate as to escape the ignorant fanaticism of their predecessors; and others, more tolerant even than the Christian princes of the time, received without distinction all men of merit who took refuge in their State, gave them employment, and recompensed them for their services. On this account philosophers and persecuted "heretics" sought an asylum among infidels, and found there the protection which Christianity did not afford,—in return for which they gave their protectors the benefits of Greek civilization.

Of all the Moslem rulers, the most distinguished for love of learning and general enlightenment was Haroun-al-Raschid, the Charlemagne of the East, contemporary and emulator of the glory of the emperor of the Franks, the hero of a hundred Arabic poems, whose dominion extended from the borders of the Indus to the heart of the Spanish peninsula. He embellished Bagdad, his capital, with schools and hospitals. His son Almamon founded the Academy of Bagdad, which became the most celebrated of the age; likewise spared no pains to draw to his court the most illustrious men of all countries. He enjoined each of his ambassadors to purchase all the writings of the philosophers and physicians that could be found, and these he required to be translated into Arabic; his interpreter, Honain, a Christian, was employed at translating for forty-

---

* See a very vigorous denial of this historical statement in *The Nineteenth Century*, October, 1894, page 555.

five years, and received, for each book rendered into Arabic, literally its weight in gold.

The eclat which the Moorish caliphs shed upon Spain from the tenth to the thirteenth century is well known. The cities of Cordova, Toledo, Seville, and Murcia possessed public libraries and academies, and students from all parts of Europe flocked to them to be instructed in arts and sciences; the library of Cordova alone embraced more than two hundred and twenty-four thousand volumes. Thus it will be seen that the dominion of mental and temporal affairs passed from the Greeks and Romans to the Saracens.

Arabian medicine constitutes one of the most interesting chapters in the history of our art. An offspring from Greek schools, it was for nearly one hundred years the foster-mother of that art, and, although it gave rise to no great discovery nor wonderful step in advance during all this period, it nevertheless kept alive all the learning of the past, and clarified rather than made it turbid. In the sixth century the Nestorians (followers of Bishop Nestor), having been driven out of Syria, settled in Persia, Mesopotamia, and Arabia, and there founded schools and other institutions such as they had had at home,—schools in which, beside the ordinary philosophic studies, medicine received a share of attention. Thus it came about that by the seventh century Arabian physicians were everywhere known and in high repute. Naturally the basis for their studies embodied the writings of Hippocrates, Galen, Oribasius, and Paul of Ægina; and the first Arabian works consisted solely of translations from the Greek, first out of their Syriac rendering, and later from the originals. Indeed, so much eminence was finally achieved by Arabian physicians that more than four hundred are known by name as authors.

The first author deserving of mention was *Bachtischua*, of Nestorian stock, celebrated in Jondisapur, director of the medical school, and later physician to Caliph El-Mansur, in Bagdad. Of his descendants several became well known in the same field.

Alkindus—this being the Latin arrangement of his Arabic name—came from a Persian family, who lived first in Basara and later at the court of the caliphs El-Monon and El-Motasin, in Bagdad. He enjoyed a very high reputation as physician, philosopher, astronomer, and mathematician, and died A.D. 873.

Mesue, the first of his name, sometimes known as Janus Damascenus, was director of the hospital in Bagdad and physician to Haroun-al-Raschid. He was born in 777, wrote extensively (since at least forty of his works have been catalogued), and died in 857 in Samarra.

Serapion the elder, also sometimes known as Janus Damascenus, and whose Arabic name was Serafiun, was born in Damascus—the exact data is not known—and died some time prior to A.D. 930. He was author of two volumes of aphorisms concerning the practice of medicine, which had at his time the greatest repute.

The most celebrated of the early Arabian physicians was Rhazes, born in the Persian province of Khorassan A.D. 850. According to the historians of his nation he was a universal genius, equally famous in music, astronomy, mathematics, chemistry, and medicine; he was surnamed "The Experienced." At the age of fifty he was one of the most distinguished professors in the Academy of Bagdad, where students came from great distances to listen to him. Chosen from among a hundred colleagues to direct the grand hospital of that city, he displayed indefatigable zeal and most scholarly learning, even to his old age and in spite of loss of sight, which overtook him at the age of eighty, when his reputation was at its height. Two years after this misfortune—*i.e.*, in 932—he died. His generosity, which was proverbial, and his compassion for the poor left him penniless at the time of his death. Some two hundred and thirty-seven monographs of his have been catalogued, though the greater number of his works are practically lost. Two treatises on medicine remain which afford excellent counsel in many respects; among other matters he advises:—

"Study carefully the antecedents of the man to whose care you propose to confide all you have most dear in this world,—that is, your life and the lives of your wife and children. If the man is dissipated, is given to frivolous pleasures, cultivates with too much zeal the arts foreign to his profession, still more if he be addicted to wine and debauchery, refrain from committing into such hands lives so precious."

His greatest publication was *Continens*—extracts compiled from all authors for his own use—divided into thirty-seven books, constituting an abridgment of the science of medicine and surgery up to his time; and, notwithstanding its imperfect state, this work was held in greatest reverence, and was a common source of knowledge among Orientals long after his day.

Haly-Abbas, a Persian by birth, flourished fifty years after Rhazes, and died A.D. 994. His *Almaleki*, in twenty volumes, constituted a quite complete system of theory and practice of medicine, which, however, was in large measure taken from Rhazes's *Continens*. It is generally regarded as the best work of any of the physicians of the Arabic Period; it is divided into three parts —a book on Health, a book on Death, and a book of Signs—and it is interesting to know that the portion devoted to midwifery and obstetrics was in the hands not only of the profession, but also of the midwives.

Avicenna—Latinized form of his Arabic name, Ebn Sina—was born in Bokhara in 980. From his earliest youth he manifested a remarkable disposition for scientific study, and it is claimed that he mastered the entire Koran at the age of ten years; also that he devoted his entire days and the greater part of his nights to research, mastering philosophy, mathematics, astronomy, and, later, medicine, which he studied at the university at Bagdad, in which city his talents were chiefly exhibited. He was received at court, loaded with favors, and elevated to the dignity of Vizier, but suddenly fell into disgrace, was deprived of property, imprisoned, and even threatened with execution. After two years, however, he was restored to liberty, and once more possessed the consideration of the public and the court, becoming the recipient of new honors. Meantime he had given himself up to intemperance, by which his previously robust constitution was undermined, and this, with excessive labor, brought about his demise at the too early age of fifty-six, in the year 1037. He was author of several books, the chief being the *Canon Medicinæ*, which remained a classic for six centuries, constituting the medical code of Asia and Saracenic Europe; no author since Galen had enjoyed so wide and extensive authority in the medical world; and in the various medical schools professors, for the most part, confined themselves to reading the *Canon* from their desks, explaining and commenting upon its text. The work was divided into five volumes, of which the first two comprised the principles of physiology, pathology, hygiene, and therapeutics, arranged to conform to the teachings of Aristotle and Galen; the third and fourth dealt with treatment; and the fifth was devoted to the preparation and composition of remedies. Avicenna appears to have surpassed in subtlety both Aristotle and Galen; he was fond of metaphysical speculation, and his works were too much filled out with subtleties of language rather than with true science. Authors of this period were fond of torturing in every way possible the writings which they undertook to edit or quote from, and, instead of devoting themselves to original research, wasted time in seeking for vague and hidden meanings. That man was most esteemed as learned who could see the greatest subtlety in some passage from one of the ancient writers; consequently, that which was obscure or unintelligible was deemed the most sublime and philosophic. A very brief study of the *Canon*, for instance, will show this, while in graphic pictures of disease the work by no means approaches those of Arctæus or Alexander of Tralles, for Avicenna too often contented himself with mentioning merely a list of symptoms without indicating in any way their progression, characters, or duration. Undoubtedly just was the criticism of an Arabian poet: "His philosophy had no sound foundation, and his medical knowledge availed him naught for the possession of personal health and long life."

Albucassis was born in Zahra, near Cordova, about the beginning of the eleventh century, and is supposed to have died A.D. 1122, at the advanced age of one hundred and one. He was author of an abridgment, or compilation, devoted to the practice of medicine, the only novelty of which is a small portion devoted to surgery, in which are described certain instruments. He says:—

"I have detailed briefly the methods of operations; I have described all necessary instruments, and I present their forms by means of drawings; in a word, I have omitted nothing of what can shed light to the profession. . . . But one of the principal reasons why it is so rare to meet a successful surgeon is that the apprenticeship of this branch is very long, and he who devotes himself to it must be versed in the science of anatomy, of which Galen has transmitted us the knowledge. . . . In fine, no one should permit himself to attempt this difficult art without having a perfect knowledge of anatomy and the action of remedies."

Not a word is said about dissections, however, from which we conclude that they were not tolerated in his time. He resorted enthusiastically to the cautery, and recommended it in spontaneous luxations and the commencement of curvature of the spine. He refers particularly to instrumental delivery and the extraction of the after-birth, and, when speaking of fractures and dislocations, he remarks: "This part of surgery has been abandoned to men of vulgar and uncultivated minds, for which reason it has fallen into undeserved contempt."

Avenzoar, born in 1113, of a Spanish family which had many illustrious scions, was instructed in medicine by his father, and ultimately achieved great celebrity through-

FIG. 6.—AVERROËS.
(From an engraving of a picture by Raffaello in the Vatican.)

out Spain and Africa; for a time he lived at the court of the Prince of Seville, loaded with honors and presents, and finally was made Vizier. Among other works he wrote a treatise on renal diseases, in which he outlined the treatment of calculus and described an operation therefor. He died in 1161.

Averroës (as he is generally known, though his Arabic name was Aben Roschd) was born A.D. 1166, in Cordova, where his father held official position. After being grounded in philosophy, mathematics, and other sciences he became a pupil in medicine under Avenzoar. The greater part of his life was passed in Seville, where he was greatly esteemed and finally knighted. In 1195 he was called to the court of the King of Spain and Morocco, in Cordova, where he received the highest honors, only, however, through some misunderstanding, to be disgraced; but he soon afterward recovered his former position and dignities. He wrote extensively not only on medicine, but on philosophy, his writings taking throughout a more or less dialectic character. He died in 1198, and from him descended a number of physicians who achieved more or less reputation.

Maimonides was born in Cordova, A.D. 1135. He early devoted himself to the Talmud, and in his extended travels visited Jerusalem; he even founded a school of philosophy in the East, which, however, had only a brief existence. He died in 1204. He ranked higher in philosophy than in medical art, and seems to have been imbued with the methods of his teacher, Averroës, and is generally regarded as a theorist rather than as a practical physician, although he wrote more or less on medical topics, and is particularly remembered for an essay upon poisons. He was about the last of the Arabians who deserves special mention.

During the period which was nearing its close at the time of the death of Maimonides, the Arabs embraced with much ardor the study of medicine, and translated into their language nearly all the treasures that had been amassed by the Greeks; indeed, the preservation of many of the great writings which would otherwise have been lost is due solely to this fact. Strange to say, however, the Arabians neglected Latin authors, and apparently possessed no knowledge of Celsus or Cœlius Aurelianus. As religious prejudices prohibited dissections, they were obliged to rely solely upon the anatomical descriptions of Galen, and succeeded in increasing the errors of the original by inaccurate translations. So far as originality of observation goes, the Arabians were in most respects behind the Greeks; nevertheless, they were the first to differentiate eruptive fevers, to which the latter paid little or no attention. The Arabian school also supplied the knowledge of purgatives, such as cassia and manna, which replaced the drastics employed by the ancients; also the mode of preparation of syrups, tinctures, distilled waters, pomades, and plasters.

While the Arabians were gradually rising by their power, intelligence, and renown, the Greeks were declining in inverse ratio; the genius, courage, and ancient virtues of the latter grew weaker and weaker, until they seemed on the verge of extinction. In the medical history of these centuries, in all Europe not under Moslem rule, there was but one man entitled to mention as an author in medicine,—viz., John Actuarius, the son of one Zacharia. He lived at the close of the thirteenth and the beginning of the fourteenth century; was employed at Constantinople, his surname being the honorary title of the court-physicians. He is more commonly known as Zacharia. Of his life we know little, save that he wrote several volumes, for the most part abridgments or commentaries on the doctrine of Galen. He laid great stress on the theory of critical days, and sustained his views by astronomical hypotheses most ingeniously combined. His was the first Greek work in which were mentioned the remedies introduced by the Arabians, yet he has not a word to say of variola, measles, spina ventosa, and other affections fully described by Arabic authors. He held remarkable views concerning the nature of man, whom he supposed to be formed by the union of two contrary substances,—the soul and the body; described somewhat elaborately an imaginary plexus of veins connected with the digestive organs, through which the animal spirits were elaborated and purified; also, and quite methodically, for his age, he explained the functions of the animal economy and the etiology of disease.

While the clouds that befogged the study of medicine in the Empire of the East thus grew heavier and heavier, we must not be blind to the melancholy spectacle concerning the provinces composing the Empire of the West. Barbarians in swarms, from the forests of Germany and Scandinavia, had swept its various portions, pillaging, destroying, and reducing to slavery its inhabitants. In southern Europe everything was changed. Each generation witnessed some new and unheard-of invader, who demanded his share of booty and renown and left a track of desolation behind him. There was a brief period of order when Charlemagne reunited under one dominion these divers races and seemed to have resuscitated the Western Empire; but no sooner was he dead than its elements, being devoid of affinity, broke apart. Former vassals, no longer restrained by the firm hand of the emperor, made common warfare against his successors and against each other, and for several ages there was nothing but a succession of wars and invasions. Feudalism gave some sort of character to this military anarchy by affording repose and, in a measure, security for those who had hitherto been trampled under foot; but learning and the sciences fell into complete neglect, and it was with great difficulty that a very small number of men found within the pale of the church a limited protection that enabled them to devote themselves to the study of medicine and ecclesiastical law. Near the end of the eleventh century, however, the enthusiasm of the crusades whetted anew the turbulent appetite of the Christian barons, and led these lords of western Europe, with their belligerent spirits, to the East, as a result of which people hitherto oppressed could breathe more freely. A few States recovered their independence; some semblance of law was established; municipal institutions were organized, and establishments consecrated to public use were founded and multiplied; finally, in the course of the thirteenth and fourteenth centuries, the cloud which covered the face of Roman Catholic Europe was in some measure dispersed, and men of talent and even genius began to appear upon the scene; everything about them being so obscure, they shone like stars in the firmament. In letters, for instance, there were Dante, Petrarch, Boccaccio; in mathematics, Leonard, of Pisa, the first in Europe to understand and employ figures and algebraic characters, although Cuvier has claimed this distinction for Gerbert, a Benedictine monk of the tenth century, who subsequently became Pope Sylvester II. At this time, although in scholastic estimation medicine, theology, and philosophy alone were fit to entertain the human mind, the natural sciences were not without occasional representatives. Roger Bacon was three centuries in advance of scientific reform, and endeavored to introduce experimental philosophy, and so fully convinced some of his auditors that they subscribed £2000 sterling to provide for the expense of his experiments; this was money most happily employed, since it made possible a number of important discoveries. It is said that Bacon knew the properties of convex and concave lenses, and was the first to conceive of the microscope and telescope; his astronomical knowledge led him to demand a reform in the calendar, which Gregory XIII carried out three centuries later; he had knowledge of gunpowder and its effects, and was, in fact, the wizard of his day; but his boldness and originality drew upon him the enmity of the church, by which he was persecuted and finally condemned to imprisonment for life upon a diet of bread and water, although he was ultimately released, in 1266, by Pope Clement IV. He wrote extensively, but only fragments of his works exist, since the friars believed them tainted with witchcraft and prevented their publication.

Before and during the time of Roger Bacon the philosophers were divided into two parties, which engaged in very unseemly and unphilosophic strife. One was termed the *Realist*, and believed, with Plato, that ideas are self-existent and independent of the mind,—in other words, veritable entities; the other, the *Nominalist*, held, with Aristotle, that general ideas are pure abstractions formed by the mind with the aid of sensations received from without, without which they could never exist,—that is, if a being could be imagined without sensibilities and the power of sensation, such being would be destitute of ideas. These two parties kept up a very active warfare, and enlisted the aid of both civil and ecclesiastical authorities, the result being persecution of each other, and that general unsatisfactory conflict into which theology and metaphysical speculation always force those who indulge in them.

Now, regarding the condition of medical affairs in the Empire of the West: Down to the seventh century, in Rome, there were court-archiaters who were attached to the retinues of the nobles, and in each large city popular archiaters formed a college charged with sanitary matters, the instruction and examination of candidates, and gratuitous services to the poor. Although there is little definite information available, it is probable that after the ruin of Alexandria much the same medical organization obtained in those provinces as continued under the Greek Empire at Constantinople. Under Arab sway we know very little of what rules or regulations governed instruction in medicine and its practice; and, so soon as one of these countries fell under the rule of the Turks, all scientific institutions seem to have decayed or been discontinued,—or, as Renouard states it: "If we may judge by what still exists to-day in this unfortunate country (Turkey), consumed by the power of ignorance and despotism, the most complete anarchy followed all older organizations."

In southern Europe, however, things had not gone on quite so badly, although at first barbarous invasion caused everywhere disorder and confusion, and the Christian States of the Western Empire yet presented after three or four centuries a chaotic condition of affairs. The ecclesiastical schools, which were under the care of the church, still pursued courses of literary and scientific instruction; in the time of Charlemagne, for instance, the colleges of the cathedrals, and even some of the monasteries, taught medicine in a very limited way under the name of physics. Thus all the liberal professions—that of medicine included—fell under the domination of the clergy, and priests, abbots, and bishops became court-physicians. The monks of Mount Cassin, of the order of St. Benoit, enjoyed for a long time a great reputation for medical skill; and among these in the tenth century was an abbot named Berthier Didier, who became Pope Victor III toward the close of the eleventh century, and one Constantine, surnamed the African. Of the ecclesiastics who from the ninth to the eleventh century were distinguished by the knowledge of medicine, there were Hugues, abbot of St. Denis, physician to the King of France; Didon, abbot of Sens; Sigoal, abbot of Epernay; Archbishop Milo, etc. Even several religious orders of women undertook, to a certain extent, the practice of medicine, and Hildegarde, who was abbess of the convent of Rupertsburg, near Bingen, is credited with having written a treatise on *Materia Medica*.

From the ninth to the thirteenth century the Jews shared with the clergy the monopoly of the healing art. Many of these studied under Arabian physicians, and, though the canons of the church forbade them to in any way minister to the ailments of Christians, they were still called upon in time of need, and even in many instances had access to the palaces of archbishops, cardinals, and popes.

The education of Christian priests and infidel practitioners embraced really very little, and consisted, for the most part, of knowledge of a few symptoms and possession of a few receipts; books were excessively rare and expensive, capable teachers lacking, and a good medical education out of the question. There was no law nor public regulation which concerned the practice of medicine, and any who desired could enter upon it; while besides the priests and the Jews—which latter stood at the top of the scale—there was a multitude of charlatans of the lowest order, such as barbers, keepers of baths, and even a few women. The morality of this vulgar herd was on a level with its knowledge. I have said the practice of medicine was not regulated by law, yet Theodoric, King of the Visigoths, enacted a statute that no physician should bleed a woman of noble birth without the assistance of a relative or domestic; that if a physician in treating a patient or dressing a wound happened to harm a gentleman he should pay a forfeit of one hundred sous, and if the patient died from the operation he should be handed over to the relatives of the deceased, who could do with him whatever they pleased; while if he crippled or caused the death of a serf, he was to be held accountable only for the loss, and compelled to supply another. This remained in force from the sixth to the twelfth century, and was made to apply chiefly to the practice of surgery, which had been abandoned to individuals of the lowest condition. The practice of internal medicine was, for the principal part, the privilege of the clergy, and it is not likely the secular power ever expected that one protected with the title of priest should be handed over to the relatives of the dead. It furthermore appears that the practice of medicine as divorced from surgery led to such irregularities in the manners and conduct of the clergy that from the twelfth century popes and councils of the church repeatedly forbade the medical art to those in holy orders or under vows; but that this prohibition was often violated is shown by the frequent reiteration of inhibitory laws. During the twelfth century the secular authority was also affected by abuses. Roger, founder of the kingdom of Sicily, one of the first Christian princes of the Middle Ages, gave special attention thereto, and in 1140 proclaimed that every one who wished to practice medicine must present himself before a magistrate and obtain authorization, under pain of imprisonment and confiscation of goods. Other sovereigns followed this example, and regulating ordinances were gradually established, which ultimately led to the institution of medical faculties and university degrees.

During the Middle Ages, in the Empire of the West, arose the School of Salernum, which became so celebrated that, like that of Alexandria, it deserves special mention. The modern city of Salerno is situated on the Neapolitan Gulf, about thirty miles southeast of the city of Naples, with a population of but a few thousand souls. The ancient city stood upon a height in the rear of the present town, where the ruins of its mediæval citadel are still to be seen. It first appeared in history 194 B.C., when a Roman colony was founded, was a municipal town of importance, and appears even at this early day to have been a health resort, since Horace informs us he had been advised to substitute its cool baths for the warm ones of Baiæ. During the stormy centuries following the downfall of the Western Empire, Salerno successively submitted to the sway of the Goths, Lombards, Franks, Saracens, and Greeks, as the vicissitudes of war compelled. Under the Lombards it became the residence of the Duke of Beneventum, and, in 1075, when taken by Robert Guiscard of Normandy, it fell to the crown of Naples, in consequence of which in the fourteenth century, the heir apparent of this kingdom took the title of Prince of Salernum.

During the Middle Ages here flourished a medical school, important not alone because of its celebrity at the time, but for its effect upon the medical history of the future. Its origin is obscure, though it has been ascribed to Charlemagne in 802; again, its founding has been held to be the work of fugitives from Alexandria when that city was captured by the Saracens, 640 A.D.; some attribute it to the Benedictine order of monks, others to Saracens, etc. The foundation by Alexandrian fugitives is probably conjectural, yet it must be admitted there is some evidence of knowledge of Arabian medicine in Salernum as early as this. Be the origin what it may, it is certain that the Benedictine monks exercised a very important influence upon this school, and there is considerable reason to think that it was really originated by them. Their monastery of Monte Casino was located about fifty miles the other side of Naples, occupying the site of an ancient temple of Apollo; the rules of the order enjoined the care of the sick and treatment by prayer, and St. Benedict himself was credited with performing miraculous cures. The rules which forbade public instruction were gradually discarded, for in the ninth century Abbot Bertharius wrote two books on the art of healing, and by the tenth century Monte Casino had acquired great reputation as a medical school, and was sought by medically-inclined monks from all quarters. A little later (1022) King Henry II, of Bavaria, Emperor of Germany, is said to have been cut for stone by St. Benedict himself, who appeared in ghostly form and operated with such skill that on awaking the royal patient found the calculus in his hand, and only the cicatrix of the wound through which it had been removed. Of course, the grateful emperor could do no less than richly endow the monastery, and bestow upon it additional privileges.

Desiderius, the Benedictine abbot from 1058 to 1086, and in the eleventh century promoted to the papal chair under the title of Victor III, was distinguished for his attainments in medicine and in music, and founded a new hospital in connection with the monastery; he also composed four books detailing the miraculous cures wrought by his patron saint. It was really within this monastery that Constantine the African, one of the most learned men and the most famous Christian physician of his time, compiled his numerous medical treatises.

About Constantine there is much of romance. He was born in Carthage in 1018 and died in 1085. He visited all the prominent schools of his day in Egypt, Bagdad, Babylon, and even India, and for thirty-nine years pursued the various branches of knowledge away from home. Returning to Carthage, misunderstood and feared, he was accused of practicing sorcery and compelled to fly to save his life. Disguised as a beggar he escaped to Salernum, which had been recently captured by Robert Guiscard, and on the recommendation of some royal visitor, who had known him at another court, he was made private secretary to Guiscard. His new duties soon became irksome, however, and he retired to a cloister to devote himself to literary labors. These, for the most part, were translations of Greek and Arabic writings, often made *verbatim* and without credit. Whatever may be said about this lack of honesty, and the barbaric nature of his Latin, credit must be given him for reviving the study of Hippocrates and Galen in France; and he is generally credited with being the first to introduce into Europe knowledge of Arabian medicine.

From Monte Casino the Benedictines at an early day spread to Salernum, where, by the middle of the tenth century, three monasteries were established, in all of which were kept holy relics. It now appears that, although there may have been some previous institution of learning at this point, and possibly even medical teachers, the real organization of a regular school of medicine was due to the Benedictines. In the annals of Naples of the middle of the ninth century the names of Salernian physicians are mentioned; and it is known that toward the close of the tenth century Archbishop Verdun visited Salernum for relief from vesical calculus, and there died.

The earliest medical writings of this school which have been preserved are found in the *Compendium Salernitanum*, discovered in manuscript form in 1837; and among the more prominent authors quoted are: Petronius, who wrote about 1035; Gariopontus, who wrote about 1040; Bartholomæus, Ferrarius, and Afflacius,—the latter a disciple of Constantius Africanus.

The preaching of Peter the Hermit, which marked the close of the eleventh century, was followed by an outburst of crusading enthusiasm that quickly converted Europe into a vast camp, and Salernum, being situated upon the highroad to the East, was benefited in no small degree and its reputation as a medical school materially enhanced; likewise its teachers gained in experience as regards military surgery. In this way it became a favorite resort for crusaders when disabled, wounded, or diseased. Robert of Normandy, son of the conqueror, returning from the Holy Land, remained here for some time with a poisoned wound in the arm, received in 1097 at the siege of Jerusalem, and it was decided it could be healed only by sucking out the poison, a process deemed dangerous to the operator. History declares that Robert's wife, daughter of Goeffrey, Earl of Conversana, being denied permission, took advantage of her husband's unconsciousness during sleep to withdraw the poison, when the wound speedily healed. At the time of the departure of Robert, hastened by the death of his brother William, John of Milan, the then chief of the medical school, presented him with the famous *Regimen Sanitatis Salerni*, said to have been composed largely for Robert's benefit. This was a Latin poem that enjoyed most unexampled popularity for many generations, and was the *vade mecum* of well-educated physicians for centuries. It is said to have passed through two hundred and forty different editions, and that more than one hundred manuscript copies are to-day to be found in various European libraries. The latest English version was published by Professor Ordronaux in 1871. A sample is here submitted:—

> "Salerno's school in conclave high unites
> To counsel England's king, and thus indites:
> If thou to health and vigor would'st attain,
> Shun mighty cares; all anger deem profane;
> From heavy suppers and much wine abstain;
> Nor trivial count it after pompous fare
> To rise from table and to take the air.
> Shun idle noonday slumbers, nor delay
> The urgent calls of nature to obey.
> These rules if thou wilt follow to the end,
> Thy life to greater length thou may'st extend."

During the twelfth and thirteenth centuries the glory of the School of Salerno reached its zenith; it was the most famous school of medicine in Europe, and was fostered by various kings. The celebrated Jew, Benjamin of Tudela, traveling from Spain to India, visited Salernum in 1164, and called it the "principal university of Christendom." Early in the twelfth century flourished Cophon, Archimatheus, and Nicholas, surnamed Præpositus, all of whom were distinguished teachers. The latter published a work known as *Antidotarium*, which was for several centuries the standard pharmacopœia, and which contained a table of weights that corresponded very closely to those of the modern apothecary. The younger Cophon, who has been confounded with his father (as both seem to have written extensively), wrote two treatises,—one on the anatomy of the hog, the other entitled *Ars Medendi*. The first is interesting as the only anatomical treatise of this school which has been preserved, and is an index of the degradation of anatomical science of that time.

The names of John and Matthew Platearius are of frequent occurrence in the records of this school, and have given rise to considerable confusion; the former is supposed to have been the husband of Trotula, a female physician, of whom I shall have more to say later.

Bernard the Provincial, who seems to have escaped the notice of most historians, wrote about 1155, and his commentary offers much interesting information concerning the therapeutics of the day; he formulated a large number of recipes to enable the sick to escape the omnipotence of the apothecaries, and recommended wine for the delicate stomachs of the more exalted of the clergy, and, inasmuch as these stomachs did not bear medicine well, he directed, in accordance with the practice of Archbishop Æfanus, that emetics should be prescribed *after* meals, when their action is less injurious and more agreeable; he advised young men and women tormented with love which they could not gratify to tie their hands behind their backs and drink water from a vessel in which a red-hot iron had been cooled. Indeed, his work is full of curious information and advice, and is not without therapeutic interest.

A name which figures largely in the history of this school is that of Magister Salernus, about which there is great uncertainty; it is not positively known whether this refers to a particular person or is a generic name covering various individuals. The name has been mentioned as that of one of the four reputed founders of the school; it is positive that there are certain treatises which bear this name, which give an appearance of authenticity to it as an individual title.

In the latter half of the twelfth century lived John of St. Paul, one of the teachers of Gilbert the Englishman; also Musandinus, who left a curious treatise on dietetics; and Urso, who wrote on the pulse and on the urine. Here in 1190 resided and studied a certain Alcadinus, from Syracuse, whose knowledge of philosophy and medicine was such that he acquired great reputation, and was made a professor; he even composed Latin medical poems.

Just at the close of this century flourished Ægidius, who

studied at Salernum, and also at Montpellier, where a school of medicine had been founded in 1180; he was physician to Philip Augustus, of France, and became professor in the University of Paris. Three treatises, all in Latin hexameter, are ascribed to him. A contemporary was Johanes Rogeriis, of Palermo, a graduate of Salernum and author of several works.

Early in the thirteenth century flourished Roger of Parma, one of the most distinguished of the alumni of this school and the earliest pioneer in modern surgery; his work on this topic, familiarly known as *Rogeriana*, enjoyed the greatest reputation in its day, and was for a long time the surgical text-book of Italy; his predilection for poultices and moist dressings in the treatment of wounds, abscesses, and ulcers became, in the hands of his successors, the distinguishing feature of the surgery of Salernum in opposition to the school at Bologna, where Hugo Di Lucca and Theodoric (his great rival) contended for the superiority of the dry treatment. Roger was also the first to use the term *seton*, and to give practical demonstration to this means of derivation.

Roland of Parma, a pupil of Roger, and a surgeon of great distinction, became professor at Bologna, and wrote a treatise on surgery, which was, for the most part, a commentary on the works of his master. The treatise of Roger and that of Roland furnished the basis for a work entitled *The Treatise of the Four Masters*, supposed to have been written about 1270, and manuscripts of which have been long known in various European libraries. It is divided into four books, displays no little surgical ability, and from its title would appear to have been the joint composition of four teachers; indeed, it was long attributed to Archimatheus, Platearius, Petro Cellus, and Afflacius, though it is now pretty generally understood to be the product of but a single pen and its author most likely a Frenchman. The ascription of authorship to four masters was probably for the purpose of increasing its weight and authority, and it constituted a reliable exposition of the surgery of Salernum in its day. It is quoted quite freely by Guy de Chauliac, who was the restorer of French surgery in the fourteenth century, and occasionally by later writers.

Another of the distinguished Salernian physicians of the thirteenth century, one highly esteemed by Frederick II, was John of Procida, who also was active in producing —if not the real author of—the massacre of the Sicilian Vespers, A.D. 1282. In a dispute concerning the question of the two Sicilies he embraced the cause of Prince Manfred, for which he was banished by Charles of Anjou, and took refuge at the court of Peter III, of Arragon, by whom he was created a baron; and he was influential in persuading the latter to assert his claim to the throne of Sicily. By various intrigues at different courts he succeeded in organizing an alliance, which betrayed its existence in this massacre, and finally resulted in the overthrow of the French in Sicily and the transfer of the island to the crown of Spain. He was author of at least two treatises devoted to medicine and philosophy.

Other writers of the School of Salernum were: a learned Jew of Agrigentum known as "Farragus," Matthew Sylvaticus, Graphæus, and Cappola. About the middle of the fifteenth century flourished Saladino, famous as an authority on materia medica.

It is of no small interest that now, for the first time in history, women began to figure somewhat prominently as writers, practitioners, and even teachers of medicine. About the middle of the eleventh century appeared a work, entitled *De Mulierium Passionibus*, attributed to the before-mentioned Trotula, wife of John Platearius, which has descended even to these days. There is nothing in the work to indicate the name or sex of the author, who is invariably spoken of in the third person; consequently Trotula's connection therewith has often been disputed.

It mentions a certain "*aqua mirabilis*," composed largely of brandy, which spirit is said to have first been employed medicinally by Thaddeus of Florence, who died in 1295; there is also an account of a patient who *wore spectacles!* The diseases of women and children are also largely dealt with. The work is undoubtedly an anonymous production of the eleventh century, disfigured by additions of a later day, and ascribed to Trotula, perhaps, because of the celebrity that attached to her; at all events, it is the earliest work ascribed to a female physician, and thus possesses special claims to interest.

Later we read of Sichelgnada, wife of Robert Guiscard and a graduate of Salernum, who endeavored to poison her step-son, Bohemond, in order to secure the succession of her own child. This infamous plot was furthered by some of the Salernian physicians, and thwarted only by the prompt action of Guiscard, who swore he would slay his wife with his own sword should the malady of Bohemond prove fatal.

Certain other female physicians of this period are mentioned, notably Abella, who, in spite of the modesty that is supposed to hedge about her sex, produced in Latin hexameter a work entitled *De Natura Seminis Hominis*. Mercuriolus, in the fifteenth century, produced treatises on the cure of wounds, pestilent fevers, and on the nails. The most celebrated of all, however, appears to have been Calenda, who lived during the reign of that notorious profligate, John II, of Naples (1414–1435), and who was particularly distinguished for her personal attractions. She graduated with great honor from the school at Salernum, and soon after, in 1423, married a nobleman of the court, which perhaps accounts for the fact that she never exercised the privilege of authorship. A little later, Marguerite, of Sicily or Naples, also a Salernian graduate, acquired an extended professional reputation, and was licensed to practice by Ladislaus, King of Poland.

Daremberg informs us that there were numerous female physicians at Salernum, much sought after because of their talents, and, moreover, highly esteemed by the professors of the school, who freely quoted the writings of their fair pupils and contemporaries; further, that they employed ointments in paralyses; fumigations, vapors, and antimony for coughs; and lotions of aloe and rose-water for swellings of the face; they combined scientific knowledge with facetious playfulness in a manner peculiar to the sex, in that they tendered unsuspecting beaux bouquets of roses doctored with powdered euphorbium, and hugely enjoyed the forced sternutations of their victims.

It will thus be seen what a wide-spread and long-continued influence the school of Salernum exerted. At first physics and philosophy were the principal branches taught, but later the other sciences were cultivated. The Emperor Frederick II united the different schools of the city into a university,—a term, however, that, as then applied, appears to have corresponded to what in the nineteenth century is understood by *corporation*. The emperor likewise published several decrees which revised the duties and privileges of practitioners of medicine and surgery in his kingdom, and, in 1224, ordered that no person should practice within the two Sicilies until examined by the faculty of the university and licensed at the royal hands; further, practitioners were compelled to devote at least one year to the study of anatomy. The faculty at this time consisted of ten professors, whose salary probably depended upon the number of pupils. A candidate for graduation was required to present proof of majority, of legitimacy of birth, and of proper duration of preliminary study, and then was examined publicly in the *Synopsis* of Galen, the *Aphorisms* of Hippocrates, or the *Canon* of Avicenna. On passing he swore to conform to all the regulations hitherto observed in medicine, to give gratuitous treatment to the poor, and to expose all apothecaries detected in adulterating drugs. A book was then placed in his hands, a ring upon his finger, and a laurel crown upon his head, when he was "dismissed with a kiss." The degree conferred was that of "*Magister*,"—the modern title of *Doctor* being at that period employed almost exclusively to designate a public teacher or professor.

But the watchfulness of King Frederick was not confined alone to the regulation of medical study within his kingdom. The number of professional visits, and the recompense therefor, were fixed by law. Every physician was compelled to visit his patients twice daily, and even once at night as well, if summoned, and for this attendance was permitted a daily fee equivalent to fourteen cents for patients within the city, while for calls without the city the largest legal charge was one dollar and thirteen cents, provided he paid his own expenses

The earlier teachings and practice of Salernum were a curious mixture of methodism, dogmatism, and superstition. The latter may be better understood when it is recalled that the practice of medicine for an extended period was confined almost exclusively to ecclesiastics, who by their very education were prone to superstition and upheld the efficacy of charms and relics, and the active intervention of saints and martyrs as well as the myrmidons of evil; hence arose many of the conflicts and absurd notions peculiar to the period. The prevalence of the doctrine of medical methodism was due to the character of the writings most accessible to students of that day,— such as those of Cœlius Aurelianus and others; and it is curious that Celsus, the most elegant of medical authors, was never popular among medical monks. The Hellenic language having almost disappeared from Italy by the sixth century, the works of the Greek authors had become a sealed book to a vast majority, even of the better educated; hence the purer sources of medical knowledge were not available. Although the school of Salernum, at a later date, prided itself upon its devotion to the "Father of Medicine," the Hippocratic writings were not known at this period; and, when Constantine the African, by the translation of Arabian works, introduced a new element into the Salernian school, he ingrafted upon its medical teaching a form of doctrine which found a congenial atmosphere, in which it throve vigorously, while, a century later, the translations of Gerard of Cremona gave a stronger impulse to the growth of Hippocratic medicine than to Hippocratic doctrine.

From the *Commentary of the Four Masters* we learn that Salernian practitioners recognized the diagnostic importance of nausea, vomiting, and the flow of blood from the ears in injuries to the head; that they resorted to the trepan for depressed fractures and the relief of intracranial extravasation; that hernia cerebri was treated by pressure and caustics; that ligatures, both above and below the opening, were applied for the treatment of wounds of the carotid arteries and jugular veins. It was advised to decline patients suffering from wounds of the heart, lungs, diaphragm, stomach, or liver, in order to avoid the disgrace of losing them; and in penetrating wounds of the intestines and in those complicated with protrusion of the wounded gut instruction was given how to envelop them in the warm abdomen of a slaughtered animal until natural color and temperature were restored, and then to insert a cannula of alder-wood into the wounded intestine, which was to be neatly closed and stitched; finally, the protrusion was to be carefully washed with warm water and returned into the abdominal cavity, enlarging the opening for this purpose. if necessary. Also was advised the extraction of diseased teeth; and the operation of lithotomy was described with considerable care. Compound fractures were to be treated with splints. On the whole, this commentary of the alleged Four Masters is the most interesting and ancient Salernian work which has been preserved, and is well worthy the attention of even modern surgeons.

Such was the school of Salernum in its prime, during the twelfth and thirteenth centuries. My readers will not have failed to note how few names have been mentioned which are prominent in medical history, and how few improvements were made in medical art by those who have been mentioned. One naturally inquires, then, what was the source of the wide-spread fame of Salerno as a school, since it was distinguished neither by notable discovery in science nor by celebrated teachers, and the predominant element was doubtless one of obstinate conservatism and unswerving devotion to ancient doctrines. Founded during the dark period of the Middle Ages, at a time when ignorance, bigotry, and superstition prevailed, it preserved, amidst the gloom that had settled upon Europe, a few rays of that intellectual light which had shown so brightly in the golden ages of Roman history. These rays, made more conspicuous by the intellectual night which they barely illumined, were a beacon for men who were groping for more light. Thus the name of Salernum became synonymous with intellectual advancement in later ages. As the parent and model of our modern university system, Salernum yet deserves, in a measure, to enjoy the esteem of a numerous scholastic offspring. At a time when priests were particularly active in passing off rudimentary knowledge for the science of healing this school began to secure all information possible from the laity for the progressive development of medicine. It began, in other words, to hold aloof and then to break away from the fetters of a fanatical church. Its decline, too, was as rapid as its career had been brilliant. One very serious blow was struck when, in 1224, Frederick II founded the University of Naples and forbade Neapolitan subjects to seek instruction at any other university. The next year a revolt in the city provoked the closure of the schools of Bologna, which were, however, opened again two years later. Within a short time the universities of Naples, Montpellier, Padua, Paris, and Bologna all entered into a contest for pre-eminence with a rivalry which was not always generous. In 1224, it is said, the latter university had no less than ten thousand students. Happily, however, the period of the Renaissance proved to be one of emancipation from the fetters of ignorance and superstition, making an appeal for liberty which the conservatism of Salernum could not brook. Roger Bacon, in England; Lanfranc and Guy de Chauliac, in France; Mondino, at Bologna, and Savonarola, at Padua, found no rivals at Salernum to successfully contest their fame. Thus this ancient school fell behind the age, and in a short time sank into a mediocrity which was scarcely brightened by the reflection of a departed glory. In 1342 Robert I renewed the decree of Frederick II, which closed all the schools in his kingdom save those of Naples, but excepted Salernum solely because of its antiquity and the traditions of his predecessors. In 1413 King Ladislaus excepted the Salernian alumni and professors from all taxes, duties, and tribute. In the middle of the fourteenth century the poet Petrarch speaks of the school as a memory of the past; but its last appearance was in 1748, when a dispute at Paris relating to the rank of physicians and surgeons was referred to Salerno's university for arbitration and final decision. In 1811 a formal decree reduced this parent of all European universities to a mere gymnasium or preparatory school; and now one may wander through the streets of the modern town and among the ruins of its ancient predecessor and seek in vain to trace some reminder of those who were illustrious during some of the most terrible ages in the world's history. No echo of tradition, no stone of ancient edifice, no library preserving precious manuscripts, not even an edition of the old Salernian regimen, in the whole city; in fact, none now so poor as to do it reverence.

CHAPTER IV.

AGE OF TRANSITION (*concluded*).—*The School of Montpellier:* Raimond Lulli, 1235-1315. John of Gaddesden, 1305- (?). Arnold of Villanova, 1234-1313. Establishment of Various Universities. Gerard of Cremona, 1187. William of Salicet, 1280. Lanfranc, 1315. Mondino, 1275-1327. Guy de Chauliac, 1300-1370.

AGE OF RENOVATION, 1400 to Present Time.—*Erudite Period*, including Fifteenth and Sixteenth Centuries. Thomas Linacre, 1461-1524. Sylvius, 1478-1555. Vesalius, 1514-1564. Columbus, 1490-1559. Eustachius, 1500-1574. Fallopius, 1523-1562. Fabricius ab Aquapendente, 1537-1619. Fabricius Hildanus, 1560-1634.

ALTHOUGH I have taken up so much time with an account of the school of Salernum, a few words must be devoted to the school of Montpellier, which was second in time and in importance among the great influences in the culture of western Europe. There was a time when to have studied there lent a special halo of glory, for, being near the sea, and in the vicinity of thermal baths, even so early as A.D. 1153 it was famous as a school of medicine; moreover, those who presided over it did not lapse unconditionally into mediæval philosophy, with its bewildering subtleties. It is said to have been founded A.D. 738, but first mention of it as a source of medical education occurs in 1137, when Bishop Adelbert II, of Mayence, visited the city to listen to its medical teachers. A faculty of philosophy was added in 1242, and one of law in 1298. Within the walls of the city sojourned both Christians and Jews, the latter being subject directly to the civil authorities, and particularly esteemed as translators. One of the most famous of the sons of Israel was Profatius Judicus, who became a rector of the faculty.

Prior to 1370, when the university became subject to the kings of France, it was under the control of the Pope; and then, as now, the school of medicine was the chief ornament of this ancient seat of learning.

One of the most illustrious and famous pupils of

Montpellier was that religious mystic and alchemistic visionary, Raimond Lull, or Lulli, a would-be transmuter of metals and seeker for the philosopher's stone. Born in 1234, at the age of thirty he began to see visions, and was thereby roused from an atheistic tendency to soon become wonderfully pious; ultimately he entered the order of Minorites, studied Arabic, and appeared as a missionary in Africa, seeking to convert the Saracens—who, however, declined the honor, and finally (in 1315) rewarded his zeal by stoning him to death. Beside works on alchemy and theology, he wrote on medical subjects, and, like all great minds of the period, passed among the common people as a sorcerer in league with the devil. Nevertheless, he was a notable figure in his age and country.

Quite celebrated became the compendium of Gilbert of England (1290), which contained the same speculative nonsense, the same polypharmacy, and the same superstition as other works of that time; what little it contained of value was taken largely from other writers. While this Gilbert, often known as Gilbertus Anglicus, was not the first English writer on practical medicine, he was the earliest whose works have been preserved.

Still more famous was John Gaddesden, physician-in-ordinary to the King of England, professor in Merton College, Oxford, who wrote the famous treatise known as *Rosa Anglica*, which appeared between 1305 and 1315. This treatise was characterized by mysticism and disgusting therapeutic measures, and tainted by medical avarice, superstition, and charlatanry. Gaddesden was, perhaps, the first to formally recommend the "laying on of hands" by the king for the cure of scrofula (first performed by Edward the Confessor—1042-1056), whence comes the ancient name for this disease,—*i.e.*, "king's evil." *

---

* A special "Service of Healing" was used in the English Church under Henry VIII, 1484-1509.

Arnold de Villeneuve (1234–1313) studied seven years at Montpellier, twenty years at Paris, visited all the universities in Italy, then went to Spain to levy on the Arabian authors. He wrote on medicine, theology, and especially on chemistry—in which art he obtained great renown both as an author and teacher. To him is due the discovery of spirit of wine, oil of turpentine, aromatic waters, besides several preparations of less note, and the introduction of chemical compounds into therapeutics. His was a very stirring life, for he traveled extensively; he became a teacher at Bologna, and physician to Peter III, of Arragon. Shortly before his demise he went to Paris, having fallen under the ban because of a declaration that papal bulls, far from being sacredly inspired, were human works, and that acts of charity were dearer to God than hecatombs, etc. He finally perished by shipwreck, but the spirit of fanaticism followed him after death, for his volumes were condemned by the Inquisition, because they commended experiments rather than mere speculations. In spite of his general honesty in accordance with the spirit of the times he inculcated deceit in medicine, and one of his declarations is: "If thou canst not find anything in the examination of the renal secretion, declare that an obstruction of the liver exists. Particularly use the word 'obstruction,' since it is not understood, and it is of great importance that people should not understand what thou sayest." He was one of the first to administer brandy, which he regarded as the elixir of life—whence the modern *Eau de Vie*.

Connected with this school, also, or well known as having studied there, were many men whose names became more or less famous—among them John Arden, who settled in London about the middle of the fourteenth century; Vinario, a contemporary of Guy de Chauliac, and the well-known surgeon and anatomist Henri de Mondeville, who was a teacher of Guy de Chauliac. But an idea of the doctrines prevalent in the medical literature of this part of the world, at this time, may be had from the fact that most writers chose titles for their works after the style of ballad singers: for instance, those describing the plague and venereal diseases were called *Flowers and Lilies of Medicine;* the *Rosa Anglica* of John Gaddesden was another example. Matters had arrived at such a pass, indeed, that men of science no longer hesitated to confess superstition and mingle it openly with deceit, to oppose the interests of the most needy, and to extort from their fellow-creatures fees in proportion to their supposed ability to pay.

In the time of Charlemagne each cathedral possessed a school in which were taught arithmetic, theology, singing, and sometimes medicine; the Episcopal College had medical teachers who gave advice and dressed wounds at the doors of the Church of Notre Dame, Paris; but when the medical profession had been divorced from the sacerdotal by councils and popes, many of these cathedral schools closed. In order to preserve the jurisdiction which they for a long time had exercised over the learned professions, many were erected into universities, and thus the clergy gave instruction in philosophy, theology, and later in medicine. During the thirteenth century arose many of the great universities in Europe, notably those of Bologna, Padua, and Naples, in Italy; of Paris, Montpellier, and Toulouse, in France; of Valencia and Tortosa, in Spain; of Oxford, in England. Pope Innocent III by papal bull guaranteed that the professors and students at Paris should be exempt from all excommunications save those which emanated directly from the Holy See; French sovereigns conferred many privileges upon the universities, and soon the members of the University of Paris formed practically a second city, with its own laws, customs, police, citizens, and magistrates. Still, however, all science belonged to the clergy, and its teachers, though removed from the cloister, were none the less Roman Catholic; so that the popes reigned over the people through the parish clergy, and over the latter by the clerical teachers and professors. Nevertheless, in all candor it must be acknowledged that these studious men, thus associated together for mutual instruction and emulation in learning, contributed, in a large measure, to elevate Christian civilization above all others, though several generations were required to secure the results calculated to make men celebrated; hence the early periods of the universities developed very few names. Many were conspicuous by their love of instruction, but not by originality of research. Men undertook expensive and wearisome voyages without encouragement or hope of reward, simply to obtain some rare manuscript or to hear some renowned professor; and they appeal to us of the nineteenth century by their devotion, if not by the results of their work.

Among the somewhat scattered and more or less eminent men of this period was Gerard, of Cremona in Lombardy, a man of great purity and studiousness, who ardnously pursued all that Latin authors could teach him, and, not being able to procure in Italy certain manuscripts which dated from the time of Ptolemy, determined to go to Toledo in search of an Arabian translation. At this time he was unacquainted with Arabic, but soon mastered it, and—armed with this powerful resource, which no other physician had possessed since the time of Constantine the African—he could not see so many Arabic works devoted to all branches of science as were gathered at the Spanish University without a desire to translate and transmit the same to his own country; hence he gave the remainder of his life to this work. He rendered into Latin the treatises of Hippocrates and Galen, of Serapion, and of all the famous Arabian authors from the time of Rhazes, including the *Canon* of Avicenna and the work on surgery by Albucassis. He died at the age of seventy-three, in 1187, at Cremona, and left all his books to the monastery of St. Lucy, within whose walls he was buried.

William of Salicet, born at Plaisance in the first years of the thirteenth century, became a professor in the University of Bologna, and later at Verona. He wrote extensively on medicine, and earned a reputation as a surgeon that preserves his fame to the present day. It is claimed that his status in medical literature depends, in large measure, upon the fact that he was, perhaps, the first to refuse slavish obedience to preceding authors, preferring, instead, to draw upon the results of personal study and experience. He died in 1280.

Lanfranc, or Lanfranchi (according to whether one prefers his French or Italian name), studied under William of Salicet. Of his early life very little is known, save that he practiced surgery in Milan at the time of the great dissension between the Guelphs and Ghibellines, and, for attaching himself to the weaker party, was exiled and forced to seek an asylum in France; he resided in Lyons for several years, and here wrote a work on minor surgery; in 1295 he went to Paris on the invitation of the faculty of medicine, opened a course on surgery which met with great success, and then published a second and larger treatise on the subject. It is said of him by Malgaigne that, less from his fault perhaps than that of his age, after his death (about 1315) surgery began to decline. From the time of Brunus, who practiced in Padua in 1250, the barbers had done the scarifying and bleeding. After the time of Lanfranchi there were others who applied leeches and often cauteries, and even the women meddled with surgery and in all operations competed with the barbers; the lay surgeons held themselves rivals to the clergy. Lanfranchi inherited from his old master, William, an aversion for them all, and often had to contend with uneducated and incompetent laymen. Clerical surgeons regarded operations as beneath their dignity; and Lanfranchi, who deplored this condition of affairs, confessed he had sometimes bled with his own hands, but had never operated for ascites, hernia, cataract, or stone.

John Pitard has descended to fame not as a writer, but as the founder of the surgical schools of St. Côme and St. Damien, which occupy so eminent a position in the surgical annals of France. In 1306 he was surgeon to the King of France, Philip le Bel, and the sworn surgeon of Chatelet. The College of St. Côme, in 1311, was only a little brotherhood of lay-surgeons, who gradually grew in importance as the result of the obstinate struggles sustained,—on the one hand, against the faculty of medicine, and, on the other, against the barber-surgeons. Malgaigne has, with great patience and clearness, shown that the importance of this body of men has been greatly exaggerated by historians; he has traced their various turns of fortune from beginning to end; I shall have occasion to consider them again farther on.

Mondino, sometimes known as Mundinus, born in 1275, became a professor in the University of Bologna, and died in 1327. He was the author of a celebrated treatise on anatomy, said to have reached twenty-five editions, and which was the first of its kind since Galen. This science had been greatly neglected; in Salernum, for instance, they were, for a long time, contented with the treatise of Copho on the anatomy of the hog, and most of the anatomical knowledge of the age was apparently derived from this source; Mondino resurrected the study and pursued it with interest and enthusiasm, though under the greatest difficulties. His works for more than two centuries, along with the writings of Galen and the Arabic authors, served for anatomical demonstration, although very incomplete,—as witness the statement:—

"Beneath the veins of the forearm we see many muscles and many large and strong cords, of which it is not necessary to attempt the anatomy on such a corpse (*i.e.*, a recent one), but on one dried in the sun for three years, as I have shown otherwise, in developing the number and the anatomy of those of the superior and inferior extremity."

On the other hand, he took the opposite course to discover and demonstrate the nerves, and advised maceration in running water. It required almost superhuman boldness to substitute demonstrations on the human cadaver for those upon swine, yet this was done by Mondino; and at the time the prejudice against dissection was so general that for more than a century after Mondino—who died in 1327—no one dared, at least publicly, to emulate his example. It was in the year 1315 that he publicly dissected the bodies of two women in Bologna. Anatomical study was further complicated at this time by certain bulls of Pope Boniface VIII, forbidding evisceration or boiling or cooking any part of the human body; these deliverances were really aimed, not against scientific investigation, but at the absurd custom introduced by the crusaders of cutting up and boiling the bodies of their relatives who died in infidel countries, in order to send them home for burial in holy ground; nevertheless, the papal injunction certainly operated to discourage and prohibit anatomical dissection, since nearly two hundred years later the University of Tübingen was obliged to apply to Pope Sixtus IV for permission to authorize dissection.

Guy de Chauliac, born in Gévaudan about 1300, was the most famous physician and surgeon in Christendom during the Arabic period. He studied at the cathedral college of Mende, which at that time was quite celebrated, and was taught medicine at Montpellier under the best masters of his day. It is probable, also, that he studied in Paris, and certain that later, in Bologna, he saw dissections made. Dissatisfaction with the writings of the ancients and the knowledge which he obtained at the schools stimulated his own powers of observation, and he became, in every respect, an original student and acquired a degree of erudition far more extended than that possessed by any of his contemporaries. He practiced in various places, longest at Lyons; and finally entered the service of Pope Clement VI, at Avignon, and probably enjoyed the same honor under Innocent V and Urban V; when the latter was made pope, in 1362, de Chauliac became his chaplain, or chapel-reader. In 1363 he published a work on surgery called *The Inventory*, upon which his fame chiefly rests, though several other volumes emanated from his pen. None knew better than he how to unite respect for the ancients with justice toward contemporaries, and he cited a large number of Greek, Arabian, and Latin authors, some of whom are now utterly unknown. The sciences, he declared, are "created by successive additions; the same man cannot lay the foundation and perfect the superstructure. We are as children carried on the neck of a giant; aided by the labors of our predecessors we see all that they have seen, and something beside." In tracing the character of a surgeon he recommends that he be "learned, expert, ingenious, bold where he is sure, timid when in doubt, avoiding bad cures and practices, being gracious to the sick, generous and compassionate, wise in prediction, chaste, sober, pitiful, and merciful; not covetous nor extortionate, but receiving moderate fees according to the circumstances of his patients, the character of the case, and his own dignity." "Never since Hippocrates," says Malgaigne, "has medicine learned a language stamped with such nobility and in such few words." Although a follower of Galen, in anatomy he insisted on the necessity of dissection, and proposed to make use of the corpses of executed criminals for this purpose. The drawings made by Henri de Mondeville were known to him; he divided abscesses into hot and cold, although among the latter he included œdema, tympanites, dropsy, scirrhus, and other conditions. In practice he was more timid, yet more active, than Lanfranchi, who never cut for stone, but left that operation to the traveling surgeons. De Chauliac described it as he had seen it performed; he opened the abdomen for dropsy, did not hesitate to attempt the radical cure of hernia, and operated for cataract. The plague which raged during the fourteenth century and depopulated the known world of one-fourth of its inhabitants, twice appeared in Avignon while Guy de Chauliac was a resident there—and he acknowledges that nothing but shame prevented him from fleeing. He remained at his post, visited the sick, and was himself attacked and left for dead. "In this frightful position he had sufficient presence of mind to follow the peculiarities of his case, analyze his own sufferings, and to give a description of them worthy of Hippocrates" (Renouard). His work soon became the surgical code of Christendom, and was commented upon and translated into all tongues, remaining for a long time a classic, and even at this day it preserves much of its interest as representing the condition of medical science at the close of the Middle Ages; moreover, its literary style was much superior to that of any of his contemporaries, all of whom wrote very barbarous Latin. He died about 1370.

With the death of de Chauliac terminates our interest, not merely in the Arabian physicians and those who were intimately connected with them, but in the so-called Arabic Period. It may be added, in passing, that the followers of Mahomet, like those of Christ, erected by the side of each of their mosques a school, and often a hospital, endowed with more or less generosity by caliphs or the wealthy, who hoped to purchase redemption and eternal happiness by such liberality.

A certain number of religious orders or communities were established during the Middle Ages to give succor to the deserving sick, the most widely known being those of St. Mary; St. Lazarus; St. John, of Jerusalem; and the

Daughters of God. To be sure, some, through the endowment of the opulent, became rich beyond all reason, and departed from their primitive purposes, and thus not only excited the covetousness of monarchs, but had even the temerity to resist their authority. This compelled, every now and again, a suppression of some order or institution— partly, perhaps, for laxity of morals, and partly because of their turbulence. Of this period it may be said that charitable zeal for the sick was never more pronounced; princes, bishops, and popes gave examples of devotion by dressing with their own hands the ulcers of lepers—and leprosy was in those days a frightful disease, having been contracted by the crusaders in the Orient, and everywhere spread as they returned, being, moreover, favored by the miserable uncleanliness which was then so common. Ignorance, dread, and fear rendered this disease worse than usual, and it was confounded with other maladies less formidable. It has been estimated that in the fifteenth century Europe harbored no less than nineteen thousand lepers; and that the disease was a great terror is manifest by the excessive caution taken against its spread: its victims were forbidden to enter cities, and on the highway were compelled to stand aside lest they should taint passers-by with their breath; even a healthy person convicted of being touched by a leper was banished from society; any infraction of these rules was punishable by death. It will thus be seen what depth of genuine humanity it required to have anything to do with one of these outcasts.

Another institution prevailed widely during these days,—namely, public baths, which were established in nearly every city and increased to such an extent that in the fifteenth century the bathers of Paris constituted a powerful brotherhood, so powerful, in fact, that Jacque Despars, physician to Charles VII, and one of the most renowned professors of the faculty, for speaking openly against the abuse of public baths, was obliged to leave the capital to avoid persecution.

A study of the general history of the Arabic Period reveals that the Arabs, previously obscure and uncivilized, emerged rapidly from the demi-savage state, and took the first rank among the polished nations of the world. During the earliest portion of this period these people were religious vandals and destructive fanatics, but later embraced with enthusiasm and persistence a study of the humanities, and endeavored to repair their early ravages by collecting the *débris* of the literary and scientific monuments of Greece; but, though they cultivated medicine with zeal and success, they added little to the Greek treasures. Later, Arabia was overrun by hordes from the deserts of Tartary, a people yet more barbarous and unknown, who established themselves in all parts of the globe then under Saracenic dominion, and by their brutal despotism degraded the Arabians to a condition approaching that from which they had emerged. This seems to have been ever the result of Turkish conquest.

Meanwhile the Greek nation, which was for so many ages at the head of civilization, gradually lost its power, virtue, courage, glory, and independence, and continued to descend, until now it exercises no influence whatever on the course of events. During the course of the Arabic Period only one Grecian physician merits mention on account of his writings, and in these there was nothing new except what he had borrowed without credit from the Saracens.

The Empire of the West.—that is, the western part of the ancient Roman Empire,—after subjugation by barbarians from Germany and Scandinavia, fell under a cloud whose darkness overwhelmed it. Its people, however, gradually received new life by commingling their blood with that of the invaders. Later they were able to repulse the Saracens who poured in upon them from Spain; then they turned their armies against each other, and wrought mutual havoc and ruin for several centuries. Again, roused by religious fanaticism, as had been the Mohammedans previously, they rushed by thousands upon the plains of Asia Minor, Syria, and Egypt, which had been for centuries occupied by the Arabs; and their adventures and enterprises, and the new and varied scenes through which they passed, gave rise among the "Francs" to some taste for poetry and works of imagination During the twelfth and thirteenth centuries governments became more stable, liberal institutions were created, the rust of ignorance gradually disappeared, and by the end of the Arabic Period there were really apparent brilliant streaks of mentality in the horizon of the nations of Europe. In this progressive movement the study of medicine shared. In the thirteenth century it was worthily represented in Italy, in Paris, and became established in Montpellier. Notwithstanding, up to this time physicians apparently only knew how to timidly follow in the track of the Arabians, and approached little, or not at all, in their studies, the purer lore of the Greeks.

THE AGE OF RENOVATION.

This Age of Renovation (extending from the commencement of the fifteenth century to the present time, according to Renouard's classification) is divided into the Erudite Period, comprising the fifteenth and sixteenth centuries, and the Reform Period, comprising the seventeenth and eighteenth centuries, and one should add, in fact, the nineteenth. In general literature this age is known as that of the Renaissance, and is one of whose beginning a great deal has been written, and so much better than I can put it in this brief work, that to general sources I should perhaps refer those who are interested in knowing how and why there came about such a tremendous change in methods and habits of thought and in acquirement of knowledge. But it is the history of medicine that at this time we particularly desire, and our minds must be, in some slight degree, prepared for the great changes to be recounted by some, with the conditions which brought about this revolution. It was truly an awakening in every department of knowledge and along every line of study; it was as if the minds of men had been dormant and lost their power of receptivity, and, after a long period of torpor, awakened in a new atmosphere amid new surroundings; as if there had burst upon them a sudden appreciation of ability to do things hitherto undreamed of, and to acquire knowledge such as hitherto had been possessed by none. Once free from the shackles imposed by authority of the past, these minds severed their Gothic bonds, and started forth in every direction with the ardor of youth and the interest of novelty, all engaging in the general enterprise of erecting from the *débris* of antique science a new temple to the mind in which to worship. While some delved among the records of the past, others sought to bind the past and present, and others, bolder yet, cut entirely loose from it, rejected all tradition, and would fain have built this temple with entirely new materials.

Now, what led to this sudden awakening? Was it chance, or the effect of certain causes which had long been operating? It has been seen that hospitals and various institutions, whose foundations were dedicated to humanity, were erected in all parts of Europe; that gradually there had come about a better social organization; that there had been a diminution of conflicts between princes and their vassals, and the relations between the two were more nearly at an equilibrium. Moreover, the invention of the compass, which rendered long voyages less dangerous and more frequent, opened up to trade regions hitherto inaccessible or unknown, and attracted interest toward commerce as a means of pecuniary gain. The telescope had been invented, and astronomy was able to seize upon some of the facts by it revealed, and thereby to make more interesting calculations concerning the motions of celestial bodies, and attain a knowledge of our solar system and its laws. Gradually the microscope shed light upon the hitherto unseen; engraving on copper had added its power of illustration to the works of the great writers as they appeared; but above all, that which brought about this condition of affairs was the discovery of the art of printing. The first attempts in this direction were made between the years 1435 and 1440, and by the united efforts of three men, whose names deserve mention so long as their art persists,— namely, Guttenberg, Faust, and Shoeffer. Thanks to them, the same information could be multiplied in manifold form and transmitted to all parts of the civilized globe. In this way intelligence and reason become triumphant; thenceforward the dominion of brute force was broken, and knowledge, because capable of dissemination, became imperishable.

At the commencement of the Erudite Period Arabic literature still predominated in medicine. Rhazes, Haly-Abbas, and Avicenna were universally invoked and explained. But a taste for Greek literature began to prevail in the universities of Italy, and was finally extended to every part of Europe, especially after the taking of Constantinople by Mahomet II, Emperor of the Turks, in 1453. This disaster, which at first bade fair to be a mortal blow to Greek literature and language, strange to say, served only to hasten their resurrection in the Occident. Constantinople having been given over to pillage at this time, most of its learned men escaped, carrying with them all manuscripts that could be seized; most of these found refuge in Italy, and enlightened protectors in the all-powerful prince of the house of Medici, in Florence, in the popes at Rome, and in Alphonso, of Arragon, King of Naples and Sicily. Everywhere these fugitives spread the knowledge of the masterpieces of Greek literature and art, and in this way a taste for books, libraries, and sound erudition was diffused, while the Greek and Latin classics were hunted up and published with great patience and ardor; thus the works of the old writers were edited, translated, commented upon, and everywhere disseminated throughout Europe.

Among those who devoted themselves to the thankless task of editing, and purifying from interpolations, the works of the classic writers was Nicholas Léonicenus, born near Vincenza in the year 1428, who studied medicine at Padua and taught it for more than sixty years at Ferrara. He possessed great vigor of mind, with purity of manners and serenity of soul, and was the first to translate directly from Greek into Latin the aphorisms of Hippocrates and portions of the writings of Galen. He combated in every way the infatuation of his contemporaries for the Arabians and their lore, and called attention to many of the errors of men who, like Pliny the naturalist, had fallen for lack of fully understanding the Greek authors they compiled. At the ripe age of ninety-six he died, regretted by all.

Thomas Linacre, of Canterbury, a contemporary of Léonicenus, though younger (1461–1524), studied first at the University of Oxford, went to Italy in 1484, and in Florence attracted the attention of Lorenzo de Medici, who made him the companion of his own children, to whom he gave the best possible advantages. In due time he returned to England, where his talents speedily won him high station, and he became physician to King Henry VIII, and later to Queen Mary. Linacre was the first Englishman, it is said, who spoke purely the language of the Romans. He translated several books of Galen that are still esteemed; and caused the founding of two chairs, one at Oxford, the other at Cambridge, whose incumbents were charged with the duty of explaining the works of Hippocrates and Galen. But he is most entitled to the gratitude of his countrymen for his influence in founding the College of London. To appreciate properly its importance and his merits, we must remember the obstacles that had to be surmounted; for at that time bishops alone had the right to accord, in their own dioceses, permission to practice medicine, and, consequently, the healing art was abandoned entirely to monks and illiterate empirics. It was well that Linacre had influence at court, else he could never have obtained the reform of such overwhelming abuses; but he triumphed in spite of powerful opposition, and secured the issue of letters patent which prohibited the practice of medicine by any one who had not received a degree in one of the two universities in the kingdom, and been examined by the President of the College of London assisted by three others. This was the achievement which gave this learned man the title of "Restorer of Medicine" in England.

Léonicenus and Linacre, who were of the early Erudite Period, also merit mention not merely because of literary talents, but because they were the first eminent physicians to embrace the study of Greek classics, and to propagate the knowledge therein contained. Subsequently others followed the same course,—too many, in fact, to be enumerated; but it was easy to follow after such leaders. From the time when men began to realize the superiority of Greek models over prolix Arabian commentaries, they were anxious to seek the light at its source, and applied themselves with avidity to the study of the originals. At this time copies of Greek authors were few in number and in a deplorable condition, owing to neglect. To rediscover them, to purify, to eliminate what was not original, to rearrange, and finally to multiply by the aid of the printing-press was an extended labor requiring great knowledge, rare sagacity, and commendable patience. One of the greatest publications in medical literature belonging to this epoch was a complete edition of the Hippocratic writings, translated into Latin by Anuce Foes,—a poor, but learned, practitioner, who lived on the products of his business as pension physician in the city of Metz,—and issued from Frankfort-on-the-Main in 1495. To this master-work Foes consecrated forty years of his life. Another treatise belonging to this same time, less important, perhaps, from a medical point of view, but nevertheless showing great erudition, was a treatise on the gymnastics of the ancients, by *Jerome Mercurialis*, a work said to be not less precious to historians than antiquarians. It was by such intense zeal and hard labor that true erudition was restored in Europe.

Following now some of the special branches of medical learning and their development, let us look first at anatomy and physiology. I have already related the salient points of the life and labors of Mondino, of whom it is said that, about the year 1315, while professor at Bologna, he dissected the bodies of two women, and shortly after published an epitome of anatomy illustrated with wood-cuts. Also has been mentioned the prohibition of anatomical study pronounced by Pope Boniface VIII, in 1300. It was only toward the close of the fifteenth and the early years of the sixteenth century that this prejudice began to abate; the popes, who then stood at the head of scientific movements, withdrew their interdictions, and the universities of Italy gave public dissections. Achillini, Benedetti, and Jacques Berenger dissected at Bologna, Padua, and Pavia, previous to the year 1500; soon afterward their example was generally followed.

Jacques Dubois, whose name was Latinized into Jacobus Sylvius, was born in 1478, in a village near Amiens; he studied in Paris, where he worked most industriously at anatomy, which later he was so successful in teaching. He was the first to arrange all the muscles of the human body, to determine their functions, and to give names to those of them which had not yet been so designated. He discovered the valves of the large veins, and was the first to study the blood-vessels by means of colored injections. He gave the same careful attention to pharmacy, and in Paris, before a large class of students, began lectures on anatomy, physiology, hygiene, pathology, and therapeutics; these he continued until the faculty, on account of jealousy, interrupted them. He then, in 1529, went to Montpellier, but returned two years later to become a member of the faculty, and once more lectured with the greatest eclat. Later yet he became a successor to Vidius in the Royal College,—a position he retained up to his death in 1555. His medical writings were extensive and marked by great accuracy, while for anatomy he did a great deal, contributing much to popularize it. He dissected a great number of animals and as many human cadavers as he could procure, the number, however, being small. Unfortunately, he subordinated all his own research to the authority of Galen, being himself among those anatomists who permitted themselves to be so far misled.

The man of genius and courage, who accepted the truth of what his eyes revealed to him, and who was the true reformer in anatomy, was Andreas Vesalius, born at Brussels, in 1514, of a family already illustrious in medicine. He studied at the University of Louvain, where he early revealed the inclinations of the anatomist, since in his leisure moments he was wont to amuse himself in dissecting small animals. Near Louvain was a place where criminals were executed; and Vesalius, having observed the body of one from which the soft parts had all been cleaned away by ravenous birds, only the bones and ligaments remaining, detached the extremities separately, and then carried off the trunk by night, thus possessing himself of his first skeleton. Attracted by the fame of Sylvius, he afterward went to Paris to become his pupil, but, not content with the lessons of his master, continued to observe for himself. On the hill Montfançon, where executions took place, he disputed with dogs and vultures for the remains of criminals, or by stealth disinterred bodies from the cemeteries at the greatest personal risk. So great was his application that his progress became rapid, and at the age of twenty he gave instruction to fellow-students; at twenty-two he became Professor of Anatomy at Padua, being appointed by the Senate of Venice; at twenty-nine he issued his great work on anat-

FIG. 7.—ANDREAS VESALIUS.
(From an old etching by Esme de Boulonois.)

omy, which showed a completeness that left far in the rear all that had hitherto been published on this subject. The following year he was called by the Emperor Charles V to the court of Madrid, then the most brilliant in Europe, where he became the first physictan, and from this time abandoned his anatomical labors.

He was the first who dared to dispute the words of Galen and point out his errors,—to ascertain that the

FIG. 8.—TITLE-PAGE OF "THE SEVEN BOOKS OF THE ANATOMY OF THE HUMAN BODY," BY ANDREAS VESALIUS, OF BRUSSELS, PHYSICIAN TO THE INVINCIBLE EMPEROR CHARLES V.

(Published in two folio volumes in Basil in 1555.)

greater part of Galen's descriptions, having been made from monkeys, did not correctly represent human anatomy. This audacity raised a crowd of vehement opponents, the least reasonable and most fanatic being his old master, Sylvius; but even these onslaughts could not conceal the truth. The minds of men generally were ripe for the revolution whose signal-fire was thus lighted, and no sooner did Vesalius appeal from the decision of Galen to observation of nature than a crowd of anatomists were ready to follow his method. He died in 1564.

One who, at Padua, had been first his pupil, then his co-laborer,—namely, Columbus, born at Cremona in 1490, —succeeded him. Columbus criticised, in some respects, the statements of his eminent predecessor, which he could better do, since he is said to have dissected fourteen bodies every year, as well as to have practiced venesection. He came so near to discovering the mystery of the circulation that it is strange how he could have missed it. He even appreciated the systole and diastole of the heart and the connection thereof with dilatation and contraction of the arteries. He knew, also, that the pulmonary veins conducted arterial blood, and that the pericardium was a shut sac. He even appreciated the lesser circulation, since he described how the blood left the right side of the heart and passed into the lungs, and came back through the veins into the left ventricle; because of this discovery, and in spite of his utter failure to appreciate the greater circulation, he has been by some regarded as entitled to the credit which is universally given to Harvey. From his position as teacher in Padua Columbus was called to Pisa, and from Pisa to Rome, where he died in 1559.

Another of the great anatomists of this period, second only in fame to Vesalius, was Eustachius, born about the beginning of the sixteenth century. He became physician to the Duke of Urbino, and in Rome a city physician and professor of anatomy, continuing to teach in the latter city

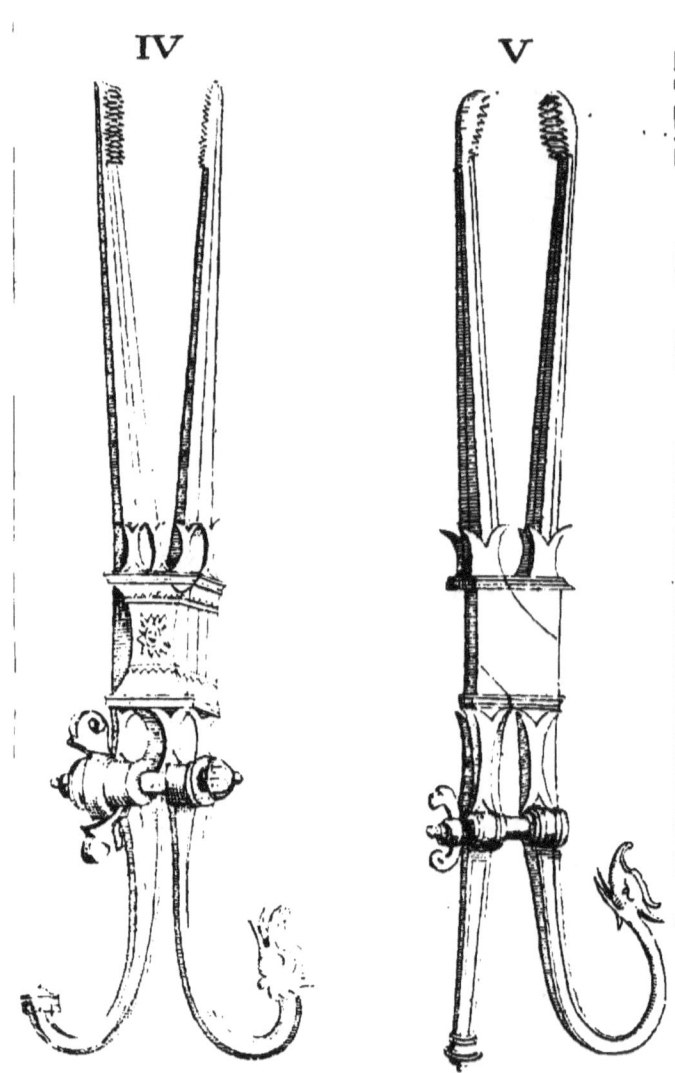

FIG. 9.—IV, FORCEPS FOR EXTRACTING BALLS. V, A DENTICULATED FORM OF FORCEPS.

(From *Opera Omnia Anatomica et Chirurgica*, by Andreas Vesalius, 1568.)

until overtaken by his final sickness. He was a defender of Galen rather than an opponent, and sought to shelter his reputation from the attacks of Vesalius. In his praise it must be said that, for his day, he was a great anatomist; his chief discoveries were in the domain of comparative anatomy. He brought to bear upon his work a knowledge of embryology which enabled him, for instance, to describe the kidneys and the teeth much more accurately than would otherwise have been possible; he noted, also the pathological changes in bodies dissected, and is brought daily to our minds as we think of the connecting channel between the pharynx and the middle ear, to which his name has been given. He died in 1574.

Fallopius, born in Modena, in 1523, was professor successively at Ferrara, Pisa, and Padua. He cultivated anatomy with the greatest ardor, and, in consequence, his name is also linked with that of Vesalius, as are those of Herophilus and Erasistratus in the history of ancient anatomy. His anatomical researches included all parts of the human body, and his name has been given to the tube through which the ovum enters the cavity of the uterus. Death overtook him in the year 1562.

Fig. 10.—GABRIEL FALLOPIUS.
(From an old etching of the sixteenth century.)

Jerome Fabricius, better known as *Fabricius ab Aquapendente*, was born in the town of the latter name, near the southern end of the Apennines, in 1537, received his early education in Padua, and studied anatomy under Fallopius, whose assistant he also was. After the death of the latter he succeeded to the professorship of anatomy, and later built, at his own expense, a large anatomical theater, in which he lectured and demonstrated to students from all parts of the world. Toward the end of his life he had accumulated a large number of specimens, and published extensively on anatomy, embryology, physiology, and surgery. Though often accredited with discovering the valves of the veins, he is not entitled to that honor, since Erasistratus, Sylvius, Vesalius, and others had previously described them, Estiennes had seen them in the azygos veins, and Canano in other veins. His true claim to glory rests upon embryological researches, which he was the first to undertake in a comparative way. In *De Formato Fœtu* he elucidated the development of the embryo and its membranes by a long list of observations on lower animals of many species. He was probably the first to describe the uterine decidua. Fabricius died in 1619.

This Fabricius must not be confused with the almost-as-renowned Fabricius Hildanus, who was born in Hilden, near Dusseldorf, in 1560. Under the German name of Wilhelm Fabry he became widely known as a surgeon, and, after traveling through France, settled in Hilden, but later moved to Cologne, where he founded an academy. His first treatise—on gangrene and sphacelus—quickly made him known, and went through eleven editions. From Cologne he went successively to Genf, Lausanne, and Polen; returned to Cologne; and finally, after several other visits, settled in Bern, where he died of gout and asthma (in 1634). His frequent changes of location were, perhaps, less the result of instability than a testimony to his reputation, inasmuch as he was invited from one place to another. He has been, with propriety, named the "German Paré," since he rendered such great service to German surgery, and was not only an expert therein, but likewise a cultivated physician and polished humanitarian; in fact he was ahead of his time, by many years, in these regards, as is shown by his recommending amputation in cases of gangrene, and his writings concerning gunshot wounds. He enjoyed a ripe experience also in obstetrics, and even instructed his wife in the obstetric art and praised her ability most highly. His most important contributions to literature were in the field of surgery, and these passed through numerous editions, while his opinions and practice are quoted even to-day.

During this epoch many modifications were introduced and improvements made in the teaching of medicine. Permanent amphitheaters were established for dissection, and chairs of anatomy created, their incumbents being paid out of the public treasury. The popes appear to have taken the initiative in this

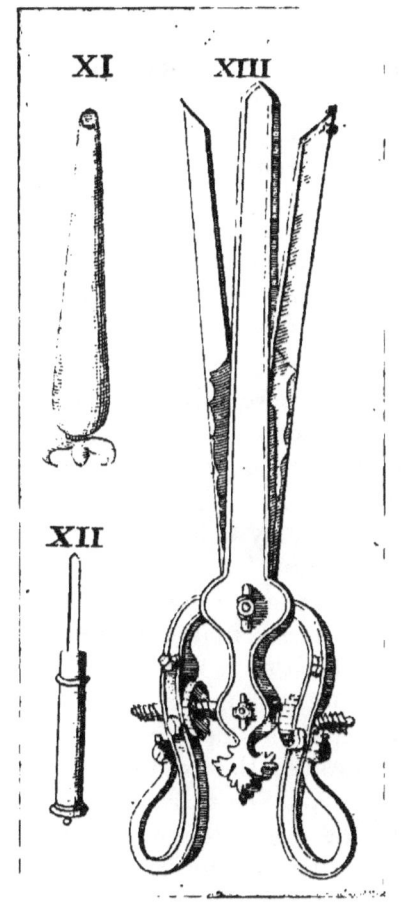

FIG. 11.—FORMS OF FORCEPS FOR ENLARGING WOUNDS.
(From *Opera Omnia Anatomica et Chirurgica*, by Andreas Vesalius, 1568.)

respect, which accounts for the great number of subjects with which Eustachius was supplied, as compared with Vesalius, who obtained only two or three in a year. Up to this time *the razor had been the sole instrument of dissection*, but was now replaced by the scalpel, which remains in use to-day. By the labors of the few men mentioned anatomy acquired a degree of perfection which it had never attained under the Greeks. Skillful artists put their labors upon paper, and plates and descriptions made from anatomical preparations represented the various parts of the human body with more fidelity than had been supposed possible. Nerves, tendons, and ligaments were no longer confused, but traced so far as possible from origin to ramifications. Ancient errors generally were corrected. It was proven that there was no bony structure in the tissue of the heart, that the partition between its cavities was not porous; and attentive examination of its valves led to the discovery of the lesser circulation by Columbus. Michael Servetus, whom John Calvin burned at the stake, was perhaps the first to note this phenomenon. He saw that the blood could not penetrate directly from the right into the left cavity of the heart, but that it was necessary for the whole fluid to pass through the lungs, where it became impregnated with the vital spirit of the atmosphere, and reached afterward the left auricle; the position of the valves in the pulmonary arteries and veins clearly confirmed his conjecture. Moreover the size of the pulmonary arteries was enormous, and disproportionate to the quantity of blood necessary for the nutrition of the lungs, which seemed to prove that this was not, as had been believed, the sole purpose of those vessels. It was about this time that Fabricius ab Aquapendente pointed out valves in veins in various parts of the body, and that Columbus and Andreas Cesalpinus explained more fully the mechanism of the lesser circulation; in fact, the former so closely approached an appreciation of the purpose of the vascular system that some have thought he really knew it, but the passages in his writings thought to sustain this opinion are not at all conclusive. He seems to have confused the action of the heart during sleep with that during the waking hours; and although he realized that the blood could not flow backward through the arteries, that the vena cava was the only vessel which permitted the entrance of blood into the heart, and though he spoke of anastomosis between arteries and veins and remarked that if a band be applied around a limb the veins swell below the ligature, he contented himself with comparing the motion of the blood with the flux and reflux of Euripus, as Aristotle had done. It is even thus that he tortured his mind in trying to reconcile two irreconcilable theories,—*i.e.*, the opinion of the ancients on the motion of the blood and recent discoveries in the anatomy of the vascular system.

CHAPTER V.

AGE OF RENOVATION (*continued*).—*Erudite Period* (*continued*): Benivieni, † 1502. Jean Fernel, 1497-1558. Porta, 1536-1615. Severino, 1580-1656. Incorporation of Brotherhood of St. Côme into the University of Paris, 1515. Ambroise Paré, 1510-1590. Guillemeau, 1550-1613. *Influence of the Occult Sciences*: Agrippa, 1486-1535. Jerome Cardan, † 1501. Paracelsus, 1493-1541. Botal, born 1530. Joubert, 1529-1583.

In the domain of pathology the Arabs had added only a very small number of observations to those contained in the works of Galen. The most interesting of these pertain to eruptive fevers. Most of their writers contented themselves with making an inventory of the acquisitions of the past, as did Guy de Chauliac, and this was about all they could do under existing circumstances; although they did not make discoveries, they prepared the way for their successors.

Two men about this time did a great deal in the direction of creating a desire for post-mortem study of cases, and in illustrating and succinctly describing symptoms.

The first of these was Benivieni, a Florentine, who died in 1502—the date of his birth being uncertain. To him, more than to any other, we owe the commencement of the study of gross pathology and pathological anatomy. He was the first to consider the knowledge that might be obtained by opening bodies for the sole purpose of ascertaining the location and cause of the diseases from which they had died. As Malgaigne remarks: "A eulogy which he merits, and which he shared with no other person, and which has not been accorded to him up to this time by the many historians of surgery who have superficially searched among these precious sources, is that he was the first who had the habit, felt the need, and set the useful example, which he transmitted to his successors, of searching in the cadaver, according to the title of his book, for the concealed causes of disease." The work referred to by Malgaigne was entitled: *Concerning Some of the Secret and Strange Causes of Disease* and was published in Florence in 1507. It is poor in quotations, but rich in original observations, which pertain especially to the etiology of disease, and gives a very concise symptomatology and history of each affection of which it treats, as well as a pathological explanation. Benivieni's observations on gall-stone, on the anatomical lesions of heart diseases, and on the conveyance of syphilis from the mother to the fœtus were original, as well as many observations concerning the presence of worms and other parasites in the body.

He did not limit himself to dissection of his own cases, but sought autopsies in the cases of others. He examined the bodies of those who had been hung, always thinking to find in them something of interest. In this regard he was followed by one already mentioned,—namely, Eustachius.

After these two the men who most cultivated pathology and anatomy in the sixteenth century were Rembert Dodoens and Marcellus Donatus. The former was born in 1517, in Mecheln, traveled extensively, was physician to Maximilian II and the Emperor Rudolph, and died in 1585. The latter lived and worked in the latter half of the sixteenth century, the dates of his birth and death being somewhat uncertain.

The next man whom we must mention is one who did a great deal for internal medicine, pathology, and anatomy. Jean Fernel, who has been surnamed "the modern Galen," was born in Clermont in 1497. Even as a boy he showed great aptitude, and very early made himself a reputation in philosophy, law, and mathematics. In 1530 he was received as doctor, with the unanimous applause of the entire faculty of Paris. He seems to have been stimulated by this only to more extended study; in fact, so hard did he work at his studies that his friends became seriously alarmed for his health, and remonstrated with him; they received for reply: "*Destiny reserves for us repose enough."* He became physician to King Henry II, of France, and in the midst of a very extensive practice undertook to collect all the medical knowledge scattered in the Greek, Arabic, and Latin works, in order to form from it a body of doctrines. His work was written with a purity and elegance of Latin that reminds one of Cicero. Throughout its pages he was philosophic, and sought to unite the apparently irreconcilable doctrines of Plato and Aristotle.

He divided medical science into three great sections,— physiology, pathology, and therapeutics. In his explanations of disease he was too often fanciful, following the speculations mainly of Galen, and making free use of the hypotheses of humors, temperaments, vital spirits, etc.; but the following statement of his would do credit to a trained pathologist of to-day: "As for myself, I shall never believe I have profound knowledge of any affection if I do not know positively, just as if I could see it with my eyes, in what part of the human body is the disease, its primitive seat, what suspicion of organic lesions constitute it, whence it proceeded, if it exists idiopathically or by sympathy, or if it be kept up by some exterior cause. He who pretends to be a rational physician must sound each of these subjects, and discern them by certain signs." The problem which he thus set himself he certainly, for his own part, considered as solved, although it was not long before his solutions were set aside and the original uncertainty reappeared.

In therapeutics he very early laid down the fundamental maxim that every disease must be combated by contrary remedies, justifying this by every species of argument, amounting to this: that every disease must be combated by its contrary because all that cures a disease is contrary to it. This was, in part, the doctrine of *"Contraria contrariis curantur,"*—the antithesis of the equally absurd sophism: *"Similia similibus curantur,"* which three hundred years later was erected into an excuse for the founda-

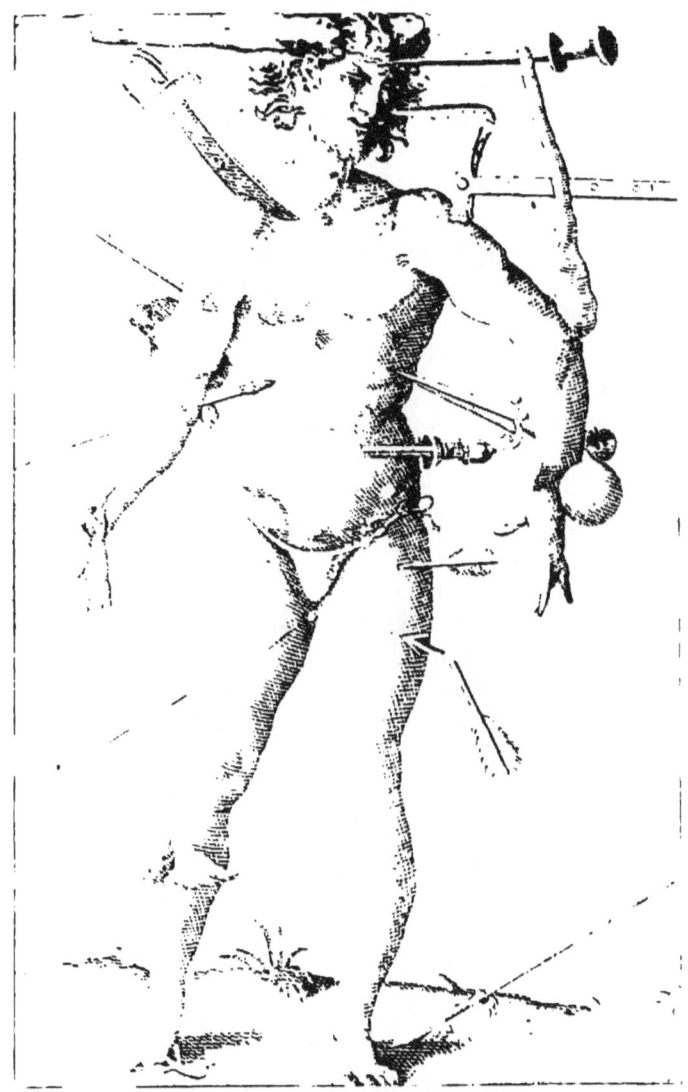

FIG. 12.—BODY SHOWING VARIOUS KINDS OF WOUNDS.
(From *Opera Omnia Anatomica et Chirurgica*, by Andreas Vesalius, 1568.)

tion of an alleged new school. There can be no doubt that Fernel rendered very great service to his time and to subsequent generations, despite the fact that his recommendations and statements were too often founded upon sophistry.

Just here we must digress for a moment to consider the status of bleeding. Hippocrates and Galen had advised to bleed largely from the arm on the affected side in pleurisy and pneumonia. That practice was gradually abandoned as Greek traditions were lost sight of, and finally the Arabs substituted for it something entirely different,—namely, pricking a vein in the foot in order to let blood flow drop by drop. Their method prevailed throughout Europe until the commencement of the sixteenth century, or about the time when Fernel appeared upon the scene. A Parisian physician named Brissot had revived the ancient (the Greek) practice during an epidemic of pleurisy, and had obtained thereby astonishing success, which he hastened to publish, commending the method employed. He thus created a great uproar in the medical world. The innovation found foes and defenders, and disputes grew warm, even to the fever point. Finally, the ancient method was generally revived, and Fernel accepted it.

Felix Plater was born in 1536, in Basel, Switzerland, and died in 1614. He had several sons who made their mark in medicine. In his large work, which preceded that of Fernel, he took perhaps the first step in an unexplored route,—namely, in the classification of disease *according to the totality of apparent symptoms.* Defective as this classification appears in our eyes, its author lived a long life as a very distinguished practitioner and professor in his native town.

Giovanni Batista Porta was born in Naples in 1536, traveled extensively in Italy, France, and Spain, and founded in 1560 an Academy of the Segreti. He was accused of magic, and was compelled to refute the charges in Rome. He died in 1615, having been one of the leading scientists of his time, and the founder of modern optics. In the first edition of his *Magia Naturalis*, published in Naples, 1587, is found the first description of the camera obscura,—of course, in a very incomplete form and without lenses.

Severino was a celebrated surgeon of Naples. He was born in 1580, in Calabria, studied in Naples, became a doctor in Salernum, and then became professor of anatomy in his native town. For a long time the victim of intrigue and of persecution by the Inquisition, he was finally driven out of Naples, but was called back by the populace. He then became the most celebrated teacher of his time, writing extensively on a variety of subjects. He died in 1656 of the plague, an epidemic of which was at that time raging in central Italy.

Arriving now at the surgery of this Age, we find that matters were more chaotic than in other departments of medicine, and for reasons which are easily given and appreciated. While, ordinarily, external diseases are more easily discerned than internal, and while in a corresponding degree they can be more satisfactorily treated ; while, in other words, external pathology has ordinarily taken precedence of internal in professional as in lay minds, this view seems to have been inverted for a time during the Middle Ages. Previous to the period now under discussion the sciences had generally declined in Europe, and surgery had fallen even lower than medicine, for the reason that medicine was in the hands of the priests, who had at that time something of a liberal education, while the practice of surgery was abandoned to a class of ignorant barbers, bathers, and bone-setters. No mechanic or artisan could take as an apprentice any youth without a certificate affirming his legitimate birth, and that he came from a family in which there were neither barbers, bath-keepers, shepherds, nor butchers. Among the men who were thus made social outcasts were those into whose hands most of the surgery of the fifteenth century fell. This was particularly the case in Germany, and other European countries were little in advance. We have seen that in France and in Italy Lanfranc and Guy de Chauliac did their best to rescue surgery from the hands of these men, but their efforts did not prevent it from being completely abandoned by the clergy, who devoted themselves to the practice of medicine.

When we come to inquire the reason for this—in other words, why an art so useful as surgery, and one which made such requirements for knowledge, sagacity, and dexterity, whose necessity was almost continually felt, particularly during these troublous times of almost constant warfare, should be so neglected by men who could best comprehend its utility and respond to its requirements—it is difficult to find a satisfactory answer. The social condition of the times sheds some light upon the question. The nations of southern Europe were socially divided at that time into the nobility, who were nearly always at war; the clergy, who monopolized learning and filled the so-called liberal professions; and, finally, the common people, who were common prey for both the other classes, and who yet had to support both without having any privileges of their own. While the practice of medicine was a clerical right, the canon of the church prohibited physicians from drawing blood, under pain of excommunication; and hence surgery, shunned by the priests, to whom it naturally belonged in connection with the practice of medicine, fell into the hands of the ignorant and vulgar, who practiced it in a purely mechanical way, without knowledge or appreciation of its possibilities. In addition to this, there was an almost total lack of detailed and precise anatomical knowledge, and but small reason to expect that the ignorant practitioners of surgery would feel the need of such knowledge. Moreover, most of the operators were itinerants, going from city to city, stopping so long as they had cases to operate upon

FIG. 13.—MODE OF EXTRACTING LEADEN BULLETS.
(From *Opera Omnia Anatomica et Chirurgica*, by Andreas Vesalius, 1568.)

or until some reverse forced them to depart. Most of these men limited themselves to one or two sorts of operations. Some operated for cataract, others for stone, others for hernia, nearly every one having a secret method which was transmitted to his posterity as a heritage.

In the history of medicine certain family names of itinerant operators have been preserved; for example, the Branca, the Norsini, in Italy, and the Colot in France.

Under such conditions there could be no such thing as the profession of the surgeon. The prejudice against dissection did not begin to abate until the thirteenth century, when a very few of the clergy dared, in a very timid manner, to perform surgical operations. Their numbers increased in the course of the fourteenth and fifteenth centuries, and in the sixteenth had become considerable. Most of the great anatomists of that period—such as Benivieni, de Carpi, Vesalius, Fallopius, and Fabricius ab Aquapendente—were great surgeons.

In due time it came about that while the clerical physicians were willing to descend to the rank of operators, the lay-surgeons aspired to the rank of doctors of medicine. This transformation took place especially in France, the only country where at that time there was a special college of surgeons—the small Brotherhood of St. Côme, already alluded to, which was always contending against the faculty on one hand and against the barber-surgeons on the other, with varying results, and which, at last, sought peace with the university and was received by it. This took place in 1515, and was the renaissance of surgery, not only for Paris, but for the whole world. By this reunion the faculty acquired authority over the barbers, who were admitted to their lectures and took courses in anatomy and surgery, gradually attaining a knowledge which entitled them to be called barber-surgeons; their rights were not curtailed, but made more difficult of procurement, for, in addition to passing their initiation for the privilege of becoming barber-surgeons, they also had to pass an examination before the physicians and the two surgeons of the king, at Chatelet, for the right to practice surgery. The surgeons, as the price of their submission to the faculty, had, beside the university privilege, a sort of supremacy over the barbers; and thus it happened that the barbers were admitted to the rank of surgeons at St. Côme, and that the surgeons of St. Côme were admitted as barber-surgeons by the faculty of medicine. In this double capacity they approached nearer the profession of medicine, from which they should never have been separated, while surgery became an art which received numerous improvements. We must now devote a little time to the consideration of at least two or three of the men who most contributed to extend and elevate it.

Among those who most contributed to make the period of which we are now speaking a glorious one, raising himself from the lowest walks of life to the attainment of the highest professional honors, is Ambroise Paré, whose name will never die while the art of surgery is taught. Paré was born about the year 1510, at Laval, of poor parents. He was an early apprentice to the provincial barber-surgeons, after which a natural ambition for improvement led him to Paris (about the year 1532), where he studied three years at the Hôtel-Dieu, and obtained the confidence of his teachers to such an extent that he sometimes operated for them. He never learned Latin, the language at that time of the books and of the schools. Paré was most fond of recalling his hospital experience; he counted it among the highest honors of his life that he should have enjoyed what he there did enjoy, and gives us to suppose that he was a favorite upon whom peculiar favors were conferred. In one of his writings, a physician of Milan having expressed astonishment at so young a man's knowledge, he remarks with pride: "But the good man did not know that I had been house-surgeon for three years at the Hôtel-Dieu de Paris." The functions of the barber apprentices in the hospital in those days were probably to make dressings and bleedings, and sometimes post-mortem examinations ordered by the chiefs, to assist the latter in their operations, and to act in case of emergency; in other words, to do about as the *internes* at present do. They probably found there a precious and rare opportunity for anatomical dissection, but it does not appear that they had regular clinical instruction.

FIG. 14.—AMBROISE PARÉ.
(From a steel engraving of the original painting in l'École de Médecine at Paris.)

Communication between master and pupil depended absolutely on the pleasure of the former.

In 1537 Paré was made surgeon to the Colonel-General of Infantry, René de Montijean, with whom he made his first campaign in Italy. (This was in the army which King Francis I assembled in Provence with which to repulse the invasion of Charles V.) He had never seen war nor recent gunshot wounds, and only knew of them by what he had read in the writings of John de Vigo. This was at a time when it was the custom of surgeons to pour boiling oil into every amputation or other wound in order to check hæmorrhage; and Paré's experience in this, his first campaign, put him in the way of his first discovery, —a discovery which will never be forgotten. He has recounted in his *Book of Arquebus Wounds* and in his great *Apology* how after the affair of Pas-de-Suze he watched the other surgeons, dreaming of nothing else but to imitate them as far as he could; how the boiling oil gave out; how his anxiety about it prevented him from sleeping; and how to his great wonder he found that the wounded who had submitted to the operation suffered more than the others. This set him to thinking, and led him, a young man without name or authority, without letters or philosophical studies, to observe, to reason, and to combat a doctrine which was universally admitted and which the highest surgical authorities of the day sustained. At that time all authors who had spoken of gunshot wounds considered them as poisonous and complicated with burns; consequently they gave the precept to cauterize with boiling oil or a red-hot iron, and at the same time to administer certain alexipharmics which should serve as internal antidotes. John de Vigo, physician to Pope Julius II, assures us that the danger of these wounds results from the round formation of the balls, from heat, and from the poisonous qualities communicated to them by the powder. His theory and the method of treatment above given had been adopted without contradiction until the day when Paré dared to utter the first protest against them.

After a campaign of three years, in which he lost his master, he returned to Paris and married. In 1543 he was in the army of Perpignan, in the service of de Rohan, grand lord of Brittany, where he gave continuous proof of his sagacity. It was after this campaign that his reputation, so well established among warriors and the nobility, inspired Sylvius with the desire of seeing him. Paré has recounted how, in a conversation which they had together, he insisted upon the then entirely new precept,

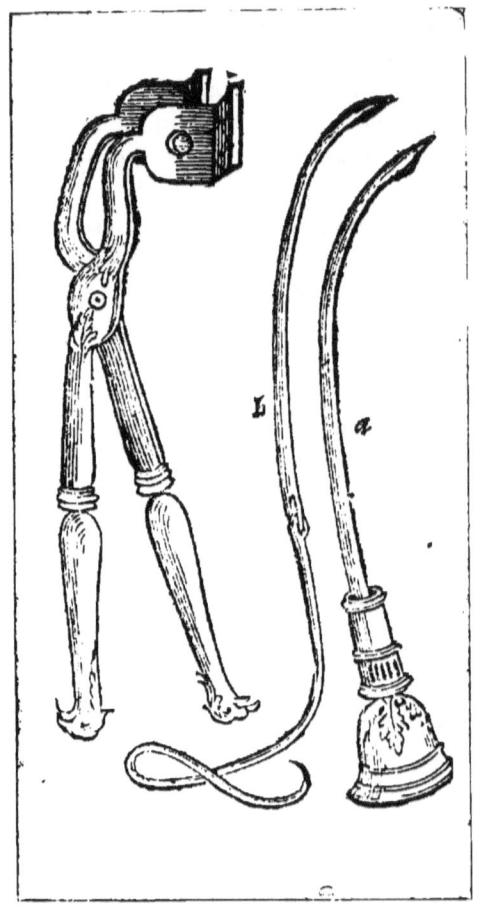

FIG. 15.—PLIERS, IRON FOR ACTUAL CAUTERY, AND SETON-NEEDLES.
(From the surgical works of Ambroise Paré, 1611.)

of which he had made many applications, that in order to extract bullets it was best to place the wounded in the position in which they were at the moment of injury.

Sylvius, then at the height of his fame, invited the young physician to dinner, and listened to him with great attention while he explained his views on gunshot wounds, which made such an impression upon the mind of the host that he besought him eagerly to write them out and make them public. Encouraged by this advice from so high a source, Paré prepared his text, illustrated it, and in the year 1545 brought out his little work, which marked in a manner so glorious the revival of French surgery. It was published by Gaulterot, the sworn bookseller of the University of Paris, and was entitled "*The Manner of Treating Wounds made by Arquebuses and other Fire-arms, and those made by Arrows, Darts, and the Like; and also by Burns made Especially by Gunpowder.* Composed by Ambroise Paré, Master Barber-Surgeon in Paris."

A few months later appeared the second edition, in which he still recommended the actual cautery in hæmorrhage; but each day he meditated upon the subject, and on one occasion discussed it with two surgeons of St. Côme, submitting to them the idea that, since ligatures were applied to veins and arteries, and to recent wounds, there was nothing to prevent their being equally applied to amputations. Both agreed with him, and opportunity soon presented itself at the siege of Damvilliers, when a gentleman had his leg crushed by a shot from the fortress. Paré made an amputation, omitting for the first time the use of the cautery, and had the happiness to save his patient, who, full of joy at having escaped the red-hot iron, said he had got clear of his leg on very good terms. This was, in truth, the actual renaissance of surgery, which had been to that time a torture, but which became thereafter a blessed art. It was a barber-surgeon who produced the double marvel. This took place in 1552.

In 1554, after other campaigns, Paré was made, without examination, Master of the College of St. Côme, and in 1559 was included among the surgeons of King Henry

II (who was killed in a tourney, in Paris, in 1559), which position he retained with Francis II and Charles IX. The latter raised him to the highest position among his surgeons, and King Henry III retained him, which caused the witty and true remark that the kings of France transferred him to their successors as a legacy of the crown.

Many anecdotes are related of Paré to show the remarkable esteem in which he was held by public and private citizens. For instance, in October, 1552, one of the most eminent generals of Charles V laid seige to the city of Metz, and the emperor came in person to join the army. Within the walls of this beleaguered city were gathered nearly all the nobility and princes of France. The city was defended by the Duke of Guise, and the besieged soldiers were at that time suffering alike from the attacks of the enemy, the results of the siege, and the rigors of a frightful winter. The duke had established two hospitals for the soldiers, and had put into requisition the barber-surgeons of the city, giving them money with which to furnish their supplies. But these surgeons were sadly incompetent against the combination of unfavorable circumstances, consequently nearly all the wounded perished, and a horrible suspicion was roused among the soldiers that they had been poisoned. Under these circumstances the duke dispatched one of his captains to the king to say that the place could hold out for ten months, and asked at the same time for fresh medicine. The king sent for Paré, gave him money, directed him to take all the medicine he thought necessary, and furnished him a letter to Marshal St. Andre, who commanded in Verdun, and who bribed an Italian captain for fifteen hundred crowns to introduce into the besieged city the celebrated surgeon. The expedition was perilous, and Paré himself would have willingly remained in Paris. But he entered Metz on the 8th of December, at midnight, without an accident. Having passed already sixteen years in war,

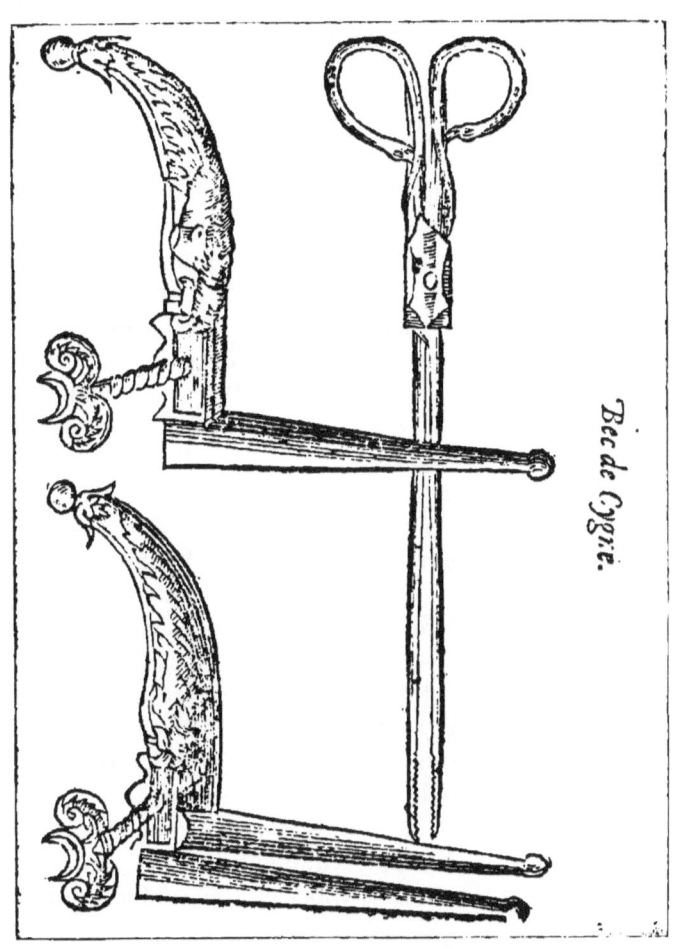

FIG. 16.—SWAN'S BEAK, USED FOR DILATING THE TRACK OF A WOUND AND EXTRACTING A FOREIGN BODY.

(From the surgical works of Ambroise Paré, 1641.)

he was known to the chiefs and common soldiers. The day after his arrival, the duke, who knew how to strike the imagination, presented him on the ramparts to all the princes, lords, and captains, who embraced and received him with clamor. By the soldiers he was received with shouts of triumph. "We shall not die," they exclaimed, "even though wounded; Paré is among us!" From this time the defense was conducted with renewed vigor, and it has been universally conceded that to the presence of this single man the city was indebted for its salvation. The siege itself was not raised until after a terrific conflict. On the very day of Paré's arrival he began to treat the leg of one of the prominent officers, who for four days had been in charge of a charlatan, and had suffered horrible tortures. The next day he decided to trephine another, who had been struck on the head by a fragment of stone, and who had been insensible for fourteen days. Both patients recovered.

The little brotherhood of surgeons of St. Côme were ready to seize on every circumstance which might redound to their advantage, and desired to have within their ranks the man who enjoyed such great renown. They, therefore, admitted him to an examination, in spite of the statute which required that the candidate should understand Latin, and in spite of opposition by the professors of the

---

DESCRIPTION OF FIG. 17.—*A*, the instrument named, on account of its figure, lizard's beak; in Latin, "*rostrum lacerti*." It is used to extract balls which have been flattened or imbedded in bone. *A* displays particularly the cannula. *B*, hinge, by means of which the lizard's beak is opened and closed as much or as little as the surgeon wishes. *C*, the rod which opens and closes the lizard's beak. When drawn upon it closes and when pushed it opens the instrument. *D, D*, dilator and mirror; in Latin, "*dilatatorium, speculum*." The instrument is somewhat roughened and dentated in order to take a firm hold of whatever it grasps. It may serve two purposes; first, to dilate and enlarge the wound so that it may be seen to the bottom, and also to make way for some instrument, as pincers or crow's beak, and to grasp more easily and withdraw the foreign body; secondly, it may itself serve to extract the foreign body,—*e.g.*, *a*, double-headed ball; *b*, a small chain; *c, c*, some pieces of mail. *E, E*, crane's beak; in Latin, "*rostrum gruinum*." *H, H*, duck's beak; in Latin, "*rostrum arserinum*." *K*, sound. *L*, ball-extractor without cannula. *M*, cannula with handle.

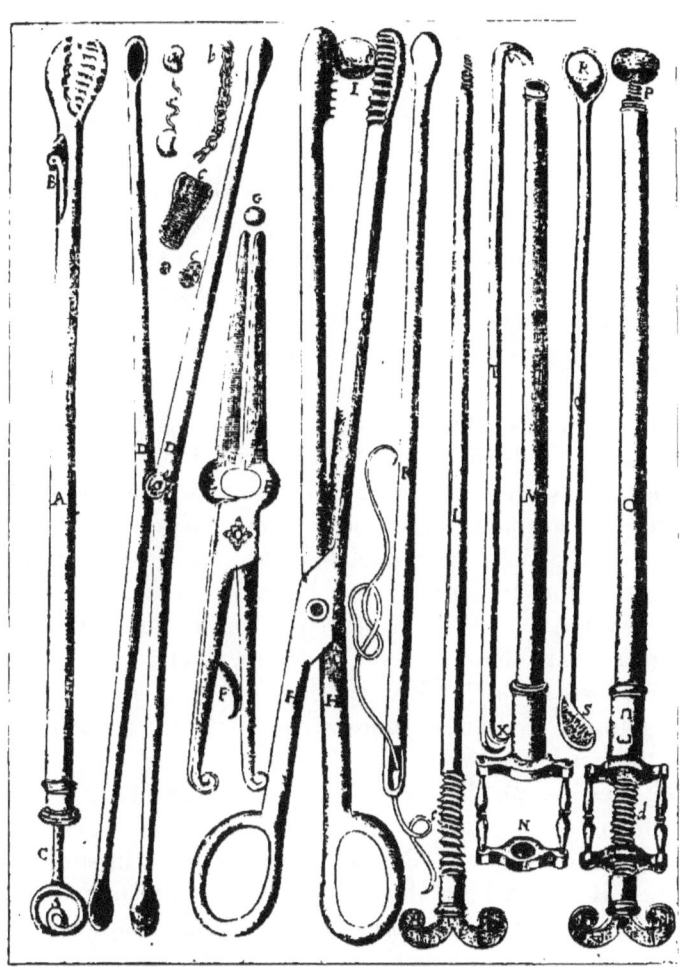

FIG. 17.—VARIOUS INSTRUMENTS FOR THE EXTRACTION OF BALLS.
(From the Works on *Chirurgerie*, by Jacques Guillemeau, chirurgeon ordinary to the King of France, 1649.)

university. They not only admitted him to all their degrees, but awarded him a reception,—a hitherto unknown honor.

Paré in his time met with a success which to-day would be pronounced extraordinary. He seemed to inspire the wounded with the utmost confidence, and to possess greatness and firmness of character in the highest degree. It is, perhaps, even more extraordinary that with so strong a character he should have so long retained favor at court. In the midst of the excitement of camps, and a very extended practice, he found time to read all that had been published on his art, and to compose himself a great number of works, enriching all branches of surgery. Instead of keeping secret his inventions, as was the custom of the time, he made them as public as possible, saying, in the preface of his large work on surgery: "For my part, I have dispensed liberally to everybody the gifts that God has conferred upon me, and I am none the worse for it; just as the light of a candle will not diminish no matter how many may come to light their torches by it."

Besides his smaller treatises, his large, collective works passed through a number of editions, and were everywhere reprinted and studied. Not only was he great in surgery, but he attained a high degree of expertness in midwifery. Among other things, he restored the forgotten practice of podalic version in cases where this procedure is necessary. He died in 1590.

The doctrine of Paré on gunshot wounds was rapidly disseminated. From 1550, Maggi, of Bologna, advocated it without giving credit to its real author, and sustained it by decisive experiments. He observed that none of the wounded felt any heat, and that the torn portions of their clothing showed no trace of fire; and he shot balls through packages of powder without setting them on fire. At the same time Lange spread this view in Germany, and Botal,

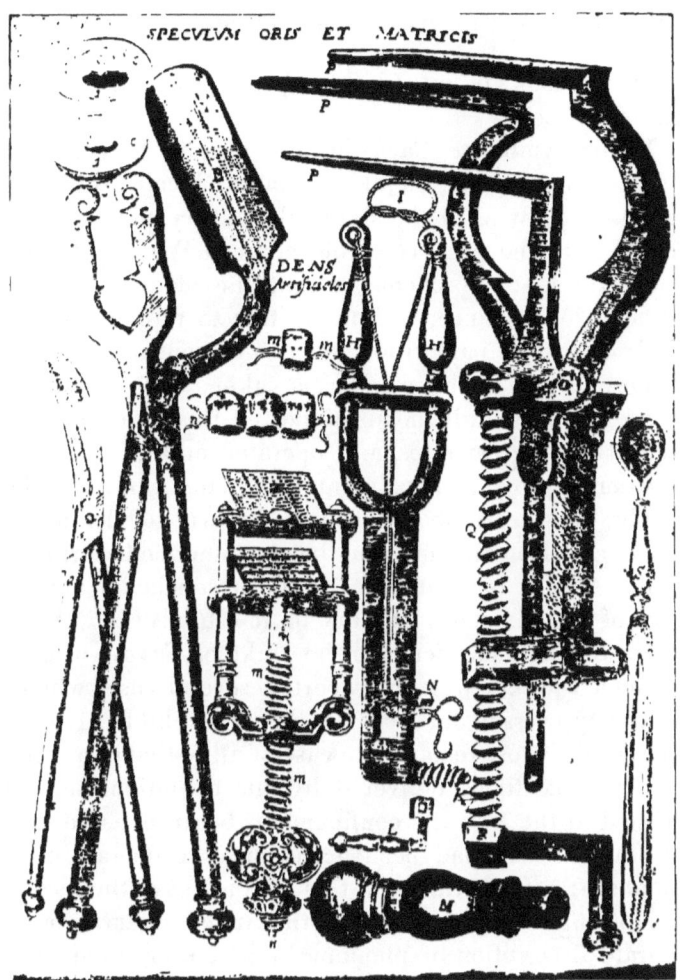

FIG. 18.—SPECULUMS FOR THE MOUTH AND WOMB, ETC.

A, A, mouth-mirror; in Latin, "*speculum oris.*" *B*, tongue-depressor, *C, C*, branches to be placed under the chin. *G, G*, instrument for retrenching elongated uvula. *O, O, O*, womb-mirror; in Latin, "*speculum matricis.*" *m, m*, artificial tooth of ivory or gold, attached by small gold threads. *n, n*, three artificial teeth joined together and attached by gold threads to the adjacent teeth on each side.

(From the *Works on Chirurgerie*, by Jacques Guillemeau, chirurgeon ordinary to the King of France, 1649.)

of Turin, took it up (withholding, however, the true author's name).

While Ambroise Paré did not disdain to act as accoucheur, it was his friend and pupil, Jacob Guillemeau (1550-1613), who, in the sixteenth century, most occupied himself with the practice of obstetrics. We owe to Guillemeau the first improvements that the moderns made in this art; for instance, the proposition to rapidly and artificially terminate parturition in cases of considerable hæmorrhage or when the woman is taken with convulsions during labor. Guillemeau supported this practice on the authority of Hippocrates, and operated on a great number of patients, proving its value and the danger of its neglect.

The Cæsarean operation was known to the ancient Greeks and Romans, but had been abandoned during the Middle Ages. It remained for the accoucheurs and surgeons of the sixteenth century to re-establish it. Among others, Rousset, physician to the Duke of Savoy, who recommended it very warmly, reported several cases where it had a happy issue for both mother and child. He even reported the most remarkable case of all,—that of a woman who was six times delivered by this operation, and who perished in the seventh confinement, because, as he states, the surgeon who had been accustomed to operate on her was absent. Unfortunately, this case is not authenticated.

Nothing shows better how the art of observation and accurate description of phenomena had progressed at the time of the revival of letters than the number of new diseases of which the authors of that period make mention. Then, for the first time did one read of whooping-cough, miliaria, scurvy, plica polonica, syphilis, and raphania. It is scarcely credible that these diseases fell upon Europe at this particular time. It is more probable that they had a more ancient existence and were not recognized.

Even to-day medical men are divided in their own opinions on the origin of syphilis, some believing that it

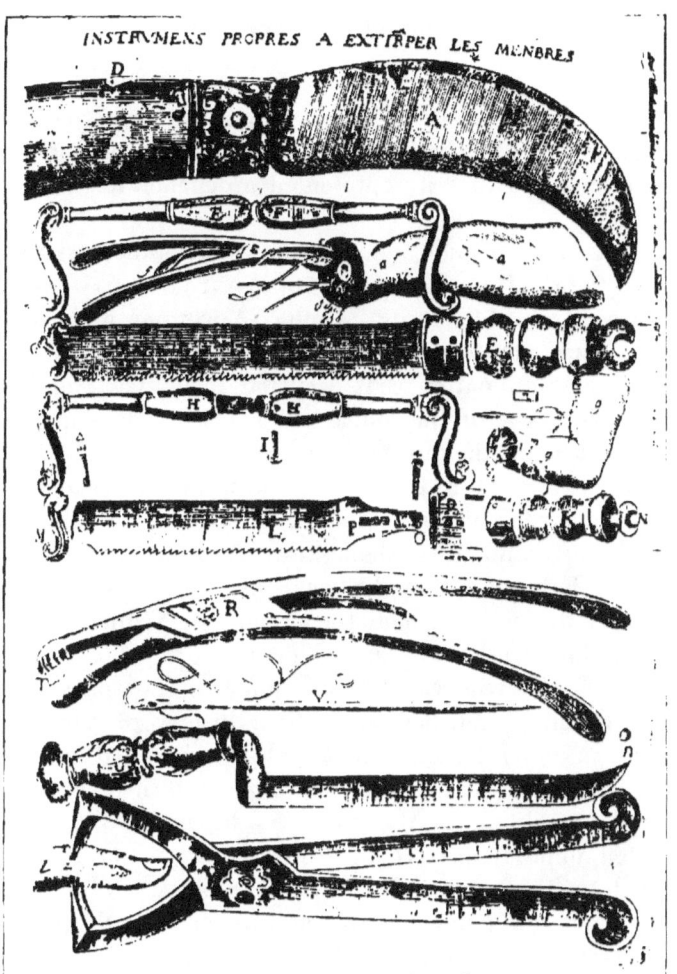

FIG. 19.—AMPUTATION INSTRUMENTS.

This plate shows knives, saws, and pliers, and also those by which hæmorrhage was arrested without use of the cautery.

(From the *Works on Chirurgerie*, by Jacques Guillemeau, chirurgeon ordinary to the King of France, 1619.)

was developed spontaneously in Europe toward the close of the sixteenth century, others that it was imported from the New World, others that it had a most ancient origin, and others yet that it represented a degenerated form of leprosy.

Certain it is that syphilis appeared almost simultaneously in all parts of Europe,—at Bologna, Halle, Brunswick, in Lombardy, Apulia, Auvergne, and so on. Léonicenus attributed this sudden outbreak to an extraordinary inundation that occurred in all parts of Italy toward the close of the fourteenth century, and supported his views with the authority of Hippocrates and Galen. Others attributed it to astrological influence; while still others regarded it as a scourge of God with which to punish men and turn them away from unbridled libertinism. Fallopius thought venereal disease was engendered by the poison which the perfidious Neapolitans had thrown into the wells from which the French drew their water. These wild views simply indicate the spirit of the age. Oviedo published in 1545 a history of the West Indies, in which he states that syphilis originated in America. He held that when Columbus returned from his second expedition to the New World, in 1496, his men enlisted under Gonsalvo de Cordova to go and fight the French, who had invaded the Kingdom of Naples, and that they communicated to the French and Neapolitans the disease which they had brought from San Domingo. Unhappily for his veracity, it is certain that syphilis broke out in Naples at least two years before the arrival of the Spanish fleet. It is equally certain that at none of the points at which Columbus touched on his return from his first expedition was there any manifestation of syphilis for years.

At this time the venereal disease, so-called, included those conditions which we now differentiate under the names of syphilis, chancroid, and gonorrhœa,—a confusion

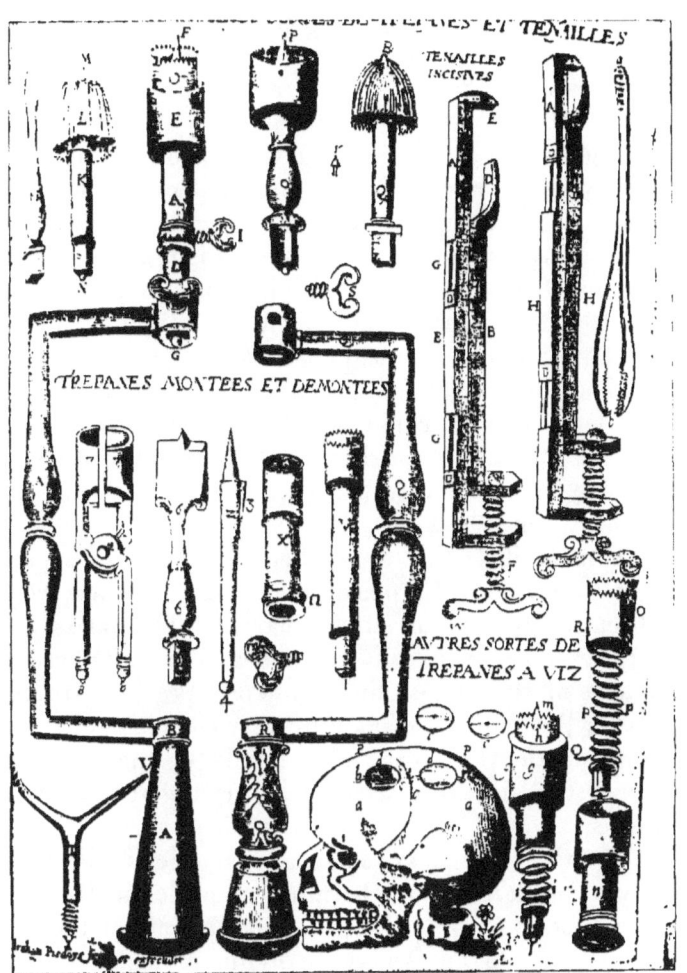

FIG. 20.—DIFFERENT FORMS OF TREPHINES AND PLIERS.
(From the *Works on Chirurgerie*, by Jacques Guillemeau, chirurgeon ordinary to the King of France, 1649.)

of diseases which persisted even up to the time of John Hunter. It is worth while to publish this fact, since writers of two or three hundred years ago may not have meant by the term "syphilis" just what we would mean to-day. Without going into this question here, it is enough to say that one who reads intelligently may see in the Sacred Scriptures unmistakable allusions to this disease. If the statements of David, as contained in the Psalms, are reliable, he was himself a serious sufferer from it. The ancient Greek and Arabian physicians make mention of lesions which could only be attributed to this disease; and the Latin satirists, like Horace and Juvenal, describe symptoms of a certain kind as being the fruit only of shameful practices.

It is most likely that the sudden appearance of syphilis in nearly all parts of Europe at about the same time, which has been regarded as so extraordinary, can be explained by the clearer distinctions physicians began to make between symptoms of this disease and those of leprosy. Arrangements for the cure of lepers were very complete, and such syphilitic patients as responded kindly to the treatment thereby established themselves in a very different category of disease.

The first writer to systematically consider venereal disease was Astruc, who was born in Languedoc in 1684 and died in 1766. He was the principal advocate of the view that syphilis had an American origin, in which view he was bitterly opposed by Sanchez, a Portuguese physician, who collected a large amount of evidence to the effect that its first ravages were observed in Italy.

Summing up this whole matter, we may agree with Jourdan, who has examined all the opinions of these writers, and who, in his treatise published in 1826, concluded that all symptoms which had been hitherto connected with syphilis had been known and described from the remotest antiquity, but were not supposed to proceed from a common source, and to be attached to the same cause, until after the close of the fifteenth century.

THE INFLUENCE OF THE OCCULT SCIENCES ON THE MEDICINE OF THIS PERIOD.

Most of the partisans of occult science were restless minds, such as are found in all ages, who chafed under the yoke of authority, and who practiced as well as deduced their lines of thought and conduct in accordance with their own ideas. Some of these men did not lack in sagacity, imagination, or audacity, but almost all of them lacked in consistency of idea and dignity of thought. Most of them lived isolated lives, apart from each other and from the rest of the world, and were, to a large extent, what we would now regard as "cranks." While they made a wide departure from accredited doctrine, they depended upon imagination rather than upon reason. This happened to be a period, however, when such men achieved great notoriety,—more so than the same class of individuals have done since their time.

Cornelius Agrippa (born in 1486) was an early promoter of occult science. He came of a noble family of Cologne, received the best education of his time, was a man of varied attainments, great inconsistency in conduct, and a caustic humor which everywhere made him enemies and prevented him from having any settled abode. He wandered from place to place, sometimes honored with the favor of the nobility and sometimes plunged into extreme misery. He early became a secretary in the court of Emperor Maximilian I, and under that monarch distinguished himself in the army by such bravery as to win him spurs as a knight. Soon disgusted with the profession of arms, he devoted himself to law and medicine, but his intemperate pen soon drew him into quarrels and persecution. At Dole he fell out with the monks;

at Paris and Turin he compromised himself with the theologians; at Metz he incurred the animosity of the Jacobins for attacking the prevailing opinion that St. Ann had three husbands. He became a vagabond and almost a beggar in Germany, England, and Switzerland, and then went to Lyons, where the mother of Francis I, who was then Queen Regent, made him her physician. He soon lost favor here, and was disgraced and banished; then he went to the Low Countries, where he was imprisoned on account of his treatise on *The Vanity of the Sciences*. Afterward he returned to Lyons, was imprisoned anew, for an old libel against his former patron, and finally died in the hospital of Grenoble, in 1535, at the age of about fifty. His treatise on *The Vanity of the Sciences* made him most trouble, and showed best both his bitterness of spirit and the extent of his learning. Herein he laid down the paradox, which was later renewed and sustained by Rousseau, that there is nothing more pernicious and injurious to common life, or more pestilential to the salvation of souls, than the arts and sciences. He founded this thesis on Scriptural authority, and supported it by profane testimony.

The conclusions which Agrippa drew were not so strange to the eyes of his contemporaries as they are to ours. Long before him, men of character and attainments, such as Pic de la Mirandola and Bessarion, had attempted to introduce the Platonic idea, that the best means of acquiring science and truth were introspective. They were, moreover, persuaded that a great number of phenomena and events have their origin in astral influences. From this system to the extravagance of the Cabal*

---

* *Cabal, or Kabbalah:* A theosophical or mystic speculative system, of Hebrew origin, which flourished from the tenth to the sixteenth century. It included a mystic theosophy and cosmogony, attributing to deity neither will, desire, nor action, but teaching that from it emanated wisdom, grace, intellect, power, beauty, firmness, and other attributes. It also ascribed hidden meanings to the sacred Hebrew writings and words. Even in the letters and forms of the sacred words

is but a step; indeed, the Christian doctrine, that events and phenomena are influenced by the direct intervention of the deity or of the devil, is but a small transposition. The cabalistic theory, summed up, was that all the events of life and all the phenomena of nature proceed from influences which gods, devils, or the stars exercised on the "archetype"—that is, on the essential spirit, or substance. He who could withdraw his spirit possessed supernatural faculties. The day and the hour of birth, according to this view, were under the domination of particular stars, and each of the principal members of the body was supposed to correspond with some planet or constellation. This is the fundamental idea underlying the pictures—which are still to be found on almanacs used by quack-medicine firms—of the individual whose interior is so completely and uncomfortably exposed, while around him are arranged the signs of the zodiac, with indications as to which part of the body is governed by each.

Occult philosophy, built upon this foundation, was divided into four branches: *theosophy*, to which a man raised himself by prayer; *magic*, or the art of controlling demons; *astrology*, or the art of reading future events by the stars; and *alchemy*, which teaches the secret of extracting the essence or the archetype of substances,— *i.e.*, virtually the secret of the philosopher's stone, by which metals were to be transmuted and then abolished.

And so the errors of science, the prejudices of the superstitious, the excitement of the religious, and the cupidity of the rich and powerful, all concurred to propagate the faults of the cabal at the close of the Middle Ages. Never were there seen so many sorcerers, astrologists, and alchemists; never were prophecies, visions, and prodigies so common. Whatever happened, it was pre-

---

the followers of the cabal pretended to find wonderful and hidden meanings; hence the modern expression "*cabalistic*." The teachings of the cabal were esoteric, of course, and inculcated mysticism and occultism in everything, but appear to have been more or less influenced by neoplatonism.

tended that it had been announced by some previous sign, or that it was a revelation of the future. This particular kind of folly persisted in Germany longer than in any other part of the world. Even Martin Luther seemed to share many of the cabalistic views, and his alleged struggle with the devil, his adventure with the inkstand, and so on, contributed much to spread them, and were, perhaps, the most prominent illustrations of their general acceptance. Surely, these *were* the Dark Ages.

Jerome Cardan was born at Pavia in 1501. His life, like that of Agrippa, was one of vicissitude and inconsistency. Being the idol of his mother and the detestation of his father produced a peculiar effect upon his character. When he began to study he made rapid progress, and at the age of twenty-two was able to discuss publicly all questions. About two years later he received his doctor's hat. He practiced medicine in various places until he was thirty-three, and was then made professor of mathematics at Milan. He occupied this position but two years, then traveled in Germany, France, and England, and returning to Italy was imprisoned for debt in Bologna, and finally obtained a pension from the pope, in Rome, where he died in 1556. He was a man of great attainments and sagacity; his literary style was dignified, and, if he had not developed such a taste for the marvelous, such inconceivable credulity and superstition, and such vanity and boasting, he would have been a remarkable character in his age. Leibnitz said of him: "Notwithstanding his faults, Cardan was a great man and, without his defects, would have been incomparable." He wrote extensively on philosophy, mathematics, and medicine. Sometimes he admitted to his writings the most absurd statements of visions, etc., and again affirmed that he had never devoted himself to cabalistic art, blamed those who practiced it, and jeered at those who believed in it. He wrote extensively on chiromancy. For his own follies and misfortunes he apologized, attributing them all to the influences of the stars.

The most colossal figure in this collection of mediæval charlatans and knaves was Paracelsus. He was born in 1493, near Zurich, of a well-to-do family, his father being a physician. He had a good preliminary education, and then visited the various universities, or rather university

FIG. 21.—PHILIP THEOPHRASTUS PARACELSUS.
(From an old engraving of a painting by Ridley.)

towns; but, instead of listening to the professors, Paracelsus associated with clever women, barbers, magicians, alchemists, and the like, from whom he acquired much information. He was led at once to the vagaries of the cabal, and, according to his own statement, he did not open a book for ten years. He neglected his studies and forgot his Latin, so that he became incapable of expressing himself in that language. From the age of twenty-five he became a hard drinker, and this habit ultimately worked his ruin. One of his disciples says of him that during the two years which he passed with him he was so inclined to drinking and debauchery that he could scarcely be seen for an hour or two without being full of wine, although that condition did not prevent him from being admired by every one as a second Æsculapius.

At this time Paracelsus was between thirty-three and thirty-five years of age, and at, apparently, the most brilliant period of his life. He had written extensively and with emphasis of his numerous cures, after the fashion of charlatans of those days,—and, unfortunately, of to-day,—and claimed to be possessed of infallible secrets against the most intractable diseases. He had just been called to Basel to the chair of physic and surgery, and crowds of curious and idle persons attended his lectures, which he gave in the vernacular, and not, as was customary in those days, in Latin. In order to strike his auditors with astonishment, he began by burning the works of Galen and Avicenna, and then reading from his own writings, breaking off from time to time into the statement: "Know, ye doctors, that my hat knows more than you; that my beard is more experienced than your academics. Greeks, Latins, Arabians, French, Italians, Jews, Christians, and Mohammedans, you must follow me; I shall not follow you, for I am your monarch, and sovereignty belongs to me." As may be imagined, his professorship was not one of long duration, and he soon had few or no listeners. In consequence of some mishaps he left Basel quite precipitately, his departure causing no such sensation as his arrival. He then resumed his nomadic life, and we find him at Alsace in 1528, at Nuremberg in 1529, at St. Galle in 1531, at Mindelheim in 1540, and in the following year at Salzburg, where he died in the hospital at the age of forty-eight.

Few men there are of whom so much good and so

much evil has been written as of Paracelsus. Few are there of whom it is to-day so hard to judge, since, if we refer to his contemporaries, they disagree completely concerning him, and if we refer to his own writings we fall into still greater chaos and have to abandon the attempt. His writings show ideas without connection, observations which contradict each other, and phrases which defy comprehension. At one moment he gives proof of admirable penetration, at the next simply abject nonsense.

That he exerted an influence upon his time is certain, but that this influence was retrograde rather than progressive seems quite likely. His exact duplicate has probably never existed since his time, and we may say that never was there another man like Aurelius Phillippus Theophrastus Paracelsus Bombastus ab Hohenheim—his full name.

Although this man was such a prominent character in his day, his name must be erased from the list of those who have contributed to the world's progress. He was simply a pretended reformer, who counted as nothing the most erudite writings, and who relied solely on his own experience. He had the most profound self-confidence, and played upon the credulity of his neighbors and victims with the toys which were furnished him by the prevalent cabalistic notions of the day. The school which he would have founded was nothing but a school of ignorance, dissipation, and boasting—a school of medical dishonesty. In a word, it was, as Renouard has said, " a school of which Thessalus, of Tralles, had been the Corypheus in antiquity, which John of Gaddesden revived in the Middle Ages, and to which Paracelsus gave a new development."

While, as has been briefly recounted, the partisans of the occult sciences strove to completely overturn the scientific edifice of antiquity, other reformers, more sensible and less daring, were content to expose its defects without attacking it in its entirety. These were, for the most part, enlightened men, and at the same time free thinkers,—friends of progress, and not of destruction. During the sixteenth century these men were few in number, but at least three or four of them deserve mention.

John Argentier was born in Piedmont, and taught in Naples, Pisa, and Turin. He did not hesitate to take issue with the theories and statements of Galen, and criticised those who adopted them too servilely. Of him it may be said that, although styled a reformer, nevertheless, he kept too near to the doctrines of those against whom he inveighed to seriously weaken their position.

Leonard Botal, also a Piedmontese, was born in 1530. First a surgeon in the French army, he later became physician to the kings Charles IX and Henri III. He was the first to recommend frequent and general bloodletting. Apparently before his time this practice was greatly restrained. He carried his views so far as to maintain that an infirm old man should be bled from two to six times a year, and that it was good custom to open the veins of healthy individuals every six months. He wrote a remarkable memoir on the cure of disease by blood-letting. It is not to be denied that he obtained some remarkable success with his copious venesections, and it must be said, in his defense, that, if he overdid it, his contemporaries did not resort to it often enough, and that his own practices were instructive to others. In his writings he united independence and energy of thought with elegance and purity of style.

Joubert (1529–1583) was Chancellor in the University of Montpellier and physician to King Henri III. He wrote a treatise on *Popular Errors*, which had an unheard-of success. In less than six months there were sold nearly five thousand copies, which, considering the times, constituted a prodigious edition. For one thing, it was written in the common tongue, and so placed within the reach of all. It was also diversified with anecdotes and jokes, some of which were not of the most delicate character; in fact, the author endeavored to atone for some of its salacity by dedicating it to Queen Marguerite. He really proposed for his main purpose a serious and useful one,—namely, that of combating prejudices which were both injurious and ridiculous. Although we may make light of Joubert's treatise, it certainly achieved a useful end by dissipating a multitude of errors, giving information to those who could scarcely get it as well from any other source. That it was full of defects is simply another form of saying that it was published in the middle of the sixteenth century.

It was during this period of which we have written that the separation of the priesthood from medicine was completed. From the sixteenth century celibacy was not obligatory on physicians in the Kingdom of France, and they no longer enjoyed ecclesiastical benefices. At this time, too, surgery, which had naturally been separated from medicine, began to approach it, the combination thus gradually brought about inuring to the benefit of all concerned. From now on, the professors of St. Côme were on the same level as the professors of the university, and enjoyed equal privileges. Institutions for instruction in medicine increased, and those which already existed were developed. Amphitheaters for dissection were open in every city in Europe. Hospitals and dispensaries were established alongside the schools, and by the various governments more attention was paid to the protection of the public from imposition, and to the amelioration of every evil affecting either public or private health.

CHAPTER VI.

AGE OF RENOVATION (*continued*).—Student-life During the Fifteenth and Sixteenth Centuries. Ceremonials Previous to Dissection.—*Reform Period:* The Seventeenth, Eighteenth, and Nineteenth Centuries. Modern Realism in Medicine and Science. Introduction of the Cell-doctrine. Discovery of the Circulation. William Harvey, 1578–1637. Malpighi, 1628–1694. Leuwenhoek, 1632–1723. Correct Doctrine of Respiration. Discovery of the Lymphatic Circulation. The Nervous System. Discovery of Cinchona. Development in Obstetric Art, in Medical Jurisprudence, in Oral Clinical Teaching. Van Helmont, 1578–1644.—*The Iatrochemical System:* Le Böe, 1614–1672. Thomas Willis, 1622–1675.

For a long time the Italian universities held the first rank; next came the French; and last the German, although all were well attended. The most famous were the medical faculties of Bologna, Pisa, Padua; then Paris, Montpellier, and, finally, Basel.

A little of what concerned the student-life of this period may not be amiss. The students chose the rector and officers of the universities, sometimes even the teachers, and assisted in determining the curriculum of study, the execution of which they watched. In some of the Scotch universities even now the students choose the rector.

The students were divided, usually according to country, into bodies denominated "*nations*" (some having special seals), which were the parents of the present student-corps in German universities. Certain representatives, known as vice-rectors, were chosen from each of these corps and constituted a so-called college of rectors which negotiated with the officials of the State, and possessed a power that was preserved until the end of the sixteenth century.

The poorer class of students passed from one school to another, supporting themselves by singing, begging, or stealing, and were sometimes guilty of great barbarities. The younger scholars, called "Schützen," were compelled to perform most menial duties for their older comrades, the "Bacchanten,"—much like the system of fagging still in vogue in English grammar-schools; and when the bacchantes were admitted to the university proper they were required to pass through an initiation, or hazing, which eclipsed anything known in these days; indeed, the antiquity of fagging may be traced back even to the philosophic schools of Athens. The habits of the traveling scholars led many of them into dissolute and vicious ways, though some attained respectable positions,—possibly even eminence. The students who were better situated financially, for the most part entered the Italian universities.

Already mention has been made of the enormous number of students congregated during this age in Bologna and in Naples. In the small University of Wettenburg there were, in 1520, only about six hundred students; in Erfurt, three hundred, and this number dwindled two years later to fifteen; in 1500 Leipzig had four hundred students; at the same time there were about seven thousand in the University of Vienna. Students and teachers migrated from one place to another, and faculties were constantly changing. Great teachers were received with great ceremony. Bitter struggles and disputes between teachers sometimes occurred; it is related of Pistorius, who died in 1523, and Pollich, deceased in 1513, that they conceived a violent enmity toward each other because of antagonistic views relative to the epidemic or contagious character of syphilis, and both ultimately left Leipzig for other schools.

Some curious customs prevailed. In teaching anatomy, while the learned teachers explained the parts as exposed, the dissections were left to barbers as being unworthy of an educated medical gentleman. While the cadavers were mainly the corpses of executed criminals, it was thought that before and after each special dissection religious ceremonies were appropriate, and such were often held; it was also believed that all who came in contact with such a corpse would be made disreputable unless it were itself first made reputable; hence the professors first read aloud a decree to that effect from the magistrate, and then, by order of the senate of the faculty, stamped upon the breast of the corpse the seal of the university. The body was next carried into the anatomical hall, and the cover of the box in which it had been transported was returned to the executioner, who remained at some distance for this purpose. If the corpse was one that had been decapitated, during these solemn ceremonies the head was placed between its legs. Finally, an entertainment with music, often furnished by itinerant actors, was given. But this folly was gradually discontinued, and by the second half of the sixteenth century public dissection was performed without recourse to such mummeries. The price of skeletons in those days was high; the University of Hiedelberg, in 1669, paid seventy-two dollars for one.

The practitioners of the sixteenth century were often quite as roving as the students and professors, though those who held positions as State physicians were bound by contract to a fixed residence for a certain time. In 1519 the State physician of Heilbronn received a salary of twenty-one dollars per year and his firewood, but could not leave the city over night without permission of the burgomaster. Medical attendants of the King of Spain were required to kneel down when they felt the king's pulse. There were few physicians who acquired wealth, although Fabricius ab Aquapendente left a fortune of two hundred thousand ducats.

The Reform Period is the name which Renouard has given to the time beginning with the commencement of the seventeenth century,—a time when the domain of natural science was daily enlarged, and when observation had enriched human knowledge with multitudes of new facts, some of which harmonized with, and some of which were in opposition to, prevailing doctrines. Men whose knowledge equaled their genius began to need a radical reform, and by such men intellectual improvement was begun by which the decrepit theories of the schools of the Middle Ages were eradicated and by which there were substituted for them others which harmonized much better with known phenomena. To the period of worship of ancient authority succeeded one characterized by a desire to shake off the yoke of the same, and men now struggled, as it were, to free themselves from the tyranny of the past. As Galileo was the torch-bearer for regeneration of the knowledge of physics, and as Kepler, and others already named, or to be named, did as much for other branches of science, so there were not lacking those who broke away from the restraint of authority in medicine, and began to beat or choose paths for themselves among the facts which experimental science furnished them.

With the approach of the seventeenth century there was evident improvement in both the social and mental status of medical men. While political humiliation and exhaustion were everywhere noted, in the field of literature it was evident that the line had advanced. What may have been the effect of thirty years of religious war, with other political struggles carried on under the hypocritical cloak of religion, may be imagined, if not fully described; the devastation of whole countries by disease, and notably by the plague,—the poverty and hunger consequent upon the ravages of perpetual war (it is stated that even so late as 1792 there were still in Saxony 535 wasted and extinct villages), to say nothing of the barbarity and immorality resulting therefrom,—all combined to make the early part of the seventeenth century a most mournful epoch. It is not strange that, with poverty, superstition and great rudeness of manners prevailed, or that trials for witchcraft and persecutions by the Jesuit Inquisition were common. That any advance should have been made under such circumstances speaks well for the progress of the human mind. That this advance was slight in Germany and central Europe is not strange, though other countries were able to quietly enlarge their scientific borders. Now it was that England, Italy, and the Netherlands, which took but little part in the warlike struggles of the century, acquired leadership in medicine, and were seconded by the French. In Great Britain, science had been fostered by various kings, and particularly by Charles II, who professed to be something of a chemist; in fact, an epidemic of scientific interest fell upon the English court.

The seventeenth century, in contrast to the idealistic sixteenth, witnessed the advent of modern realism in almost all departments of thought. Medicine furnished the first example in what we are accustomed to-day to speak of as the exact method; hence, the century is of great importance, in that physicists and chemists began to be original, instead of mere followers of the past. The most notable feature of medicine was the promulgation of three medical systems: the pietistically colored Paracelsism of Van Helmont; the chemical system of Sylvius; and the iatrochemical system of the physicist and mechanician, Borelli. This period is, moreover, illumined by the life of one great practitioner, whose name will be imperishable in the history of our art,—namely, Sydenham.

The principal tendency of the time was toward skepticism, which had begun in the preceding century with Montaigne, and was continued by Charron, under the patronage of Queen Marguerite of Navarre; it was the fundamental idea of Pierre Bayle, the author of the great dictionary. Opposed thereto was the supernatural philosophy, or the theosophic, cabalistic, or mystic. The leading exponent of the latter was Boehme, who was a business colleague of the celebrated "Meistersinger," Hans Sachs, in Germany, and of Blaise Pascal and his contemporary,

Malebranche, in France. The doctrine of Lord Bacon, Lord Verulam (1561-1626), a man who showed himself as exalted in mind as he was mean in personal traits, was of great importance Bacon is a landmark in history as the defender and eulogist of modern realism,—*i.e.*, of inductive philosophy. While personally contributing but little to the advance of science, he taught a great method; as Gruen says, he was the philosopher of patents and profit; he recognized the compass, the art of printing, and gunpowder as great inventions, but placed little value on the discovery of Copernicus, having little comprehension of mathematics. Hobbes and Locke went farther into realistic philosophy, and the latter was an exponent later of pure empiricism

In the seventeenth century, also, zoölogy and botany were largely extended. In it lived Swammerdam (1637 -1680), famous as a naturalist, physiologist, linguist, poet, and *savant;* there were others, also, whose names are better known in the history of collateral science than in medicine, and who left conclusive demonstrations in accordance with their theories, and made daily use of the microscope, simple as it then was. The term "cell" had been introduced by Hooke in 1667, and Malpighi and Grew were the founders of the cell-doctrine. The astronomical laws discovered by Copernicus changed the course of the world's thought; and now appeared the brilliant Kepler (1571-1630), and Galileo (1564- 1642), the defender of the Copernican system, and the persecuted discoverer of the law of falling bodies, of the thermometer, the telescope, and the movements of Jupiter; also, Sir Isaac Newton (1642-1727). whose discovery of the laws of gravitation in 1665 marked an era in the history of science. This century, too, gave birth to Römer, who in 1675 calculated the velocity of light; Huyghens (1627-1693). who discovered the polarization of light and the satellites of Saturn ; James Gregory, who in 1663 made a reflecting telescope with a metallic concave mirror; Torricelli, who in 1643 measured the weight of the air; Gascoigne, who invented the micrometer in 1639; and Napier, who invented logarithms in 1700.

Now chemistry, having ceased to be alchemy, began to don the dignity of a science *per se*, and it may be claimed that medicine derived no slight benefit therefrom. Scientific societies and journals arose at this period, and were all of good service to medicine in their way. The church scented danger to the faith in everything which related to natural science, and founded certain secret associations, especially in Italy; the Accademia Degli Lyncei, so called from its seal, which bore the image of a fox or lynx, founded in Rome in 1603, was one of these. Counter-societies, or, rather, societies with opposite purposes, were also started, and the original and private so-called Invisible Society, which was originated mainly by Milton, in 1645, and remodeled by Charles II in 1662, is now the flourishing Royal Society. In France the Academy was founded in 1665 by Colbert, but developed its first real activity thirty-five years later.

Those who to-day are so familiar with the course of the circulation of the blood through the arteries and veins find it difficult to understand how the recognition of this phenomenon could have been so long delayed; it seems so simple, yet to the ancients it was perfectly incomprehensible! Although every one had recognized that blood would flow from an incision, few stopped to reason thereupon. From time immemorial it had been supposed that the veins had their origin in the liver, and were the only vessels which contained blood, since the arteries were always found empty after death; the latter were held to contain only air or spirit. The circulation was supposed to leave and return to the liver through the venous canals by undulating movements similar to those of the waves of the ocean; and this was the doctrine of the

Asclepiadæ, and probably of Erasistratus. Galen modified this view by showing that the arteries contained blood; he knew it was poured into the right cavities of the heart by the great veins, but he believed that only a small quantity passed from the right ventricle into the lungs, and that the major portion reached the left ventricle by passing through pores in the inner ventricular septum. This opinion was uncontested until the middle of the sixteenth century.

Then the theologian, Michael Servetus, who, in 1553, perished as the victim of Calvin's jealousy, denied the passage of the blood through this septum, contending that it was returned from the lungs to the left side of the heart by the pulmonary veins. This was a happy thought, and a great step toward the truth. Soon after Columbus demonstrated anatomically that the conjecture of Servetus was plausible, by showing the function and real use of the valves of the heart. Cesalpinus came still nearer to the truth, and explained, as did Columbus, the course of the circulation through the lungs, but he opined that blood and vital spirits passed from the arteries into the veins during sleep, because at that time there was swelling of the latter and diminution of the pulse. Valves in the veins were known, and it had been shown that ligature of an artery in the living animal stopped the flow below it, while if a vein were tied there was shrinkage above the ligature, and swelling below it. Such was the state of science at the beginning of the seventeenth century; there remained, practically, but one step to take,—to find the true course of the blood.

William Harvey was born in Folkestone, Kent, in 1578 and died in London in 1637. He first studied at Cambridge, entering at the age of fifteen; subsequently traveled in France, Germany, and Italy, remaining in Padua from 1599 to 1602, in order to hear the lectures of Fabricius ab Aquapendente. With the title of "Doctor" he returned and settled in London and soon became a member of the College of Medicine, of which he was made a regent in 1613; in time he became physician to James I. and, on the demise of this sovereign, to Charles I; to the latter he dedicated his chief work. During the civil war he was driven from place to place, and, finally, to Oxford, where he surrendered himself to the Parliamentary troops, after which he again resided in London with his brothers,

Fig. 22.—William Harvey, M.D.

who had become rich. Modesty led him to decline the high distinction of President of the College of Physicians, and he lived a quiet and retired life, occupied with his studies and, in his later years, investigations in mathematics. Soon after 1613 he began, through his lectures, to make known the doctrine of the circulation of the blood; but he did not publish the results of his researches until 1628, after submitting them to fifteen years of proofs and counterproofs of every kind. So bitter was the opposition of his contemporaries to the new doctrine that he at one time lost a part of his practice, and was even held to be demented. It is characteristic of the fate of new truths, as well as of that age of dominant authority, that his first publication— *Concerning the Motions of the Heart and the Blood*—was unable to pass censorship in England, and therefore appeared in a foreign country (Frankfort, in 1628) when he was fifty years old; but his second treatise on the same subject, in reply to Riolan, a professor in the Faculty of Paris, was published in Cambridge in 1649.

"So much care and circumspection in search for truth, so much modesty and firmness in its demonstration, so much clearness and method in the development of his ideas," says Renouard, "should have prepossessed every one in favor of the theory of Harvey; but, on the contrary, it caused a general stupefaction in the medical world, and gave rise to great opposition."

This theory, which to-day appears so natural that we conceive with difficulty why it was not sooner discovered, was nothing less than a revolution in physiology; it excited a tremendous controversy that continued more than twenty-five years, and in which mingled every one possessed of any pretension to knowledge of anatomy or physiology; even naturalists and philosophers took part in the dispute. René Descartes was the first to declare in its favor and to support it by experiment; John Walæus (Jan de Wale), the celebrated Professor of Anatomy in the University of Leyden, confirmed it by new observations; finally Plempius, of Louvain, for a time one of the most fiery of opponents, succumbed to the truth, and in 1652 passed publicly to the ranks of Harvey's followers—no small triumph!

During these long debates Harvey remained always dignified and firm, although the early attacks rendered him unduly sensitive regarding others which he anticipated.

About his only answer to the arguments adduced against him, was to add new proofs and new experiments to those already published. The only one of his adversaries who obtained a direct response was Riolan, who possessed immense influence among his contemporaries as a man of attainments; Riolan combated with equal violence and obstinacy the other great discovery of the age,—viz., the circulation of the lymph. Harvey ultimately, however, had the satisfaction of seeing his theory universally adopted. But his services were not limited to this one discovery. He made most interesting observations on generation, both in man and in animals; on midwifery; and on the structure and diseases of the uterus.

The intermediary system and bond of union between the arteries and veins, so very essential, yet up to this time unknown, was discovered by the great Malpighi, who was born in 1628 near Bologna, became professor in its university, and discovered in the lungs and mesentery of frogs, in 1661, the capillary circulation. He first described the corpuscles of the blood in 1665; he also discovered the lung-cells, as well as the cutaneous glands, certain portions of the kidney, and the pigmentary layer of the skin, named after him (*rete Malpighi*), which later furnished the first explanation of the difference of color in different races.

In 1690 Leuwenhoek (1632–1723), who had been making observations on the larvæ of frogs and other small animals, was able to see with his improved microscope the movements of the blood in the small vessels, and gave the important testimony of his observations. In 1687 Cowper saw the passage of the arterial into the venous current in the mesentery of a cat. The capillary connection between the two vascular systems was first demonstrated by Marchetti, but was best shown by Ruysch (1638–1731), professor at Amsterdam, the famous inventor of minute injections, who greatly advanced anatomy by the formation of collections, one of which was brought into Russia by Peter the Great at an expense of about seventy-five thousand dollars. The Russian transporters of the collection, however, drank the alcohol in which many of the preparations were preserved, and a portion of the specimens was thus ruined.

Further illustration and amplification of Harvey's views came from various sources; the last, perhaps, from Nicholaus Steno (1638–1686), who was first a professor in Copenhagen, then a bishop and peripatetic converter of heretics. Steno first proved the heart to be a muscle that contracts actively and expels the blood. The duct that bears his name was discovered during his residence in Leyden or at Amsterdam. His name is written also "Stenson."

While ancient anatomists were able to describe in a general way the form of the lungs, their location, consistency, the ring-like structure of the trachea, and the first division of the bronchi, they did not go farther, but blindly accepted the prevalent theory that the bronchial tubes anastomosed with the terminal pulmonary veins, and that in this way atmospheric fluid was conveyed from the respiratory organs into the heart. On such vague and erroneous data was constructed the theory that the air was drawn into the lungs by the heat of the heart, which was the reservoir of the vital spirits; that in penetrating through the smaller tubes it was rarefied, its thinnest part passing into the heart, where it served as material for the formation of the vital spirit, its grosser part being exhaled. In other words, respiration was supposed to have two purposes, one to refresh the lungs, which, being porous and inflammable, would otherwise take fire from the heart, or focus of animal heat; the other to furnish the pneuma, or ether, which was employed by the heart in the formation of animal spirits. Harvey's discovery upset all this, in great measure.

Next it was shown that pulmonary veins carried nothing to the heart except blood. And now, during this Reform Period, the purpose of the movements of the chest was better studied, for Borelli, Helvetius, and Haller made many experiments, as the result of which it was determined that during inspiration the thorax is enlarged in all directions, and during expiration partly collapsed by relaxation of muscles, and that there never is any empty space between the lungs and the sides of the chest; further, that air is drawn into the chest by the tendency of all gases or fluids to maintain an equilibrium, or, in other words, because Nature abhors a vacuum. This being settled, various pneumatic theories were adopted and abandoned, all of which had subsequently to give way before a knowledge of what really occurs. The truth was conceived of by Mayow in 1668. It had been noticed that blood which appeared black in issuing from the veins, became red in contact with the air, and direct observation proved a similar change of color to take place during its passage from the pulmonary veins during life. Goodwin, opening the thorax of a frog, was the first to see this, and Hessenfratz filled a silk bladder with venous blood, and, plunging it into an atmosphere of oxygen, saw the blood change from black to red. In this way and by the later labors of Bichat and Lavoisier were clearly established the mechanism and the purpose of the function of respiration.

The discovery of the lymphatic vessels and their purpose was scarcely less remarkable than that of the circulation, though marked by less eclat because it was not the work of one man, but a matter of slow development. Herophilus and Erasistratus had seen white vessels connected with the lymph-nodes in the mesentery of animals, and supposed them to be arteries full of air. Galen disputed this, for he believed that the intestinal chyle was carried by the veins of the mesentery into the liver. In 1563 Eustachius described the thoracic duct in the horse. In 1622 Aselli, Professor of Anatomy at Milan, discovered the lacteal vessels in a dog which had been killed immediately after partaking of food; having pricked one of these by mistake, he saw a white fluid issue from it. Repeating the experiment, he became certain that the white threads were vessels which drew the chyle from the intestines. He observed the valves with which they are supplied, and supposed these vessels all met in the pancreas and continued on into the liver. In 1647 Pecquet, while still a student at Montpellier, discovered the lymph-reservoir, or *receptaculum chyli*, and the canal which leads from it (the thoracic duct), which he followed to its termination in the left subclavian vein. Having ligated the duct, he saw it swell below and become empty above the ligature. He studied the courses of the lacteals, and convinced himself that they all entered into the common reservoir. This discovery gave the last blow to the ancient theory which attributed to the liver the function of blood-making, and confirmed the doctrine of Harvey. Strangely enough, the latter united with Riolan in opposing the discovery of Pecquet and denying its significance. From this time the lymphatic vessels and glands became objects of common interest and were investigated by many anatomists,—by Bartholin, Ruysch, the Hunters, Hewson, and, above all, by Mascagni, who was the first to give a graphic description of the whole lymphatic apparatus.

The ancients confounded, under the name "neuron," nerves, tendons, ligaments, and membranes; even Aristotle regarded the brain as an inert mass devoid of sensation, and supposed the nerves to originate in the heart. Rufus. of Ephesus, remarked that Herophilus distinguished three sorts of nerves,—the first serving for sensation and motion and proceeding from the brain and spinal marrow, the second and third serving to unite bones and muscles. Galen also shared in this error, but, nevertheless, described the brain-membranes and the difference between white and gray matter; he supposed the cerebrum to be the seat of the soul and origin of sensory nerves, and that the cerebellum gave rise to nerves of motion; the pulsation of the cerebrum exposed was held to be a sort of brain respiration. Galen came very near recognizing the distinction between nerves and tendons, but nevertheless confused them. The anatomists of the sixteenth century described certain portions of the nervous system with more exactness than did Galen, but not with such positiveness as to prevent Cesalpinus from renewing the Aristotelian theory that the heart was the origin of sensation and the seat of the soul. Nearly two centuries later Baglivi advanced a theory which referred vital movement to the heart and the dura mater.

The progress which accrued to comparative anatomy and physiology, and the experiments which were made on animals, during this period, shed a great deal of light upon the nervous system. The researches of Vieussens, Haller, Meckel, Vicq d'Azyr, Scarpa, Soemmering, and others had already rendered it manifest that the brain was the organ of sensation and voluntary motion, and Bichat had proposed to divide the nervous system into cerebro-spinal and sympathetic branches.

Now, too, Kepler discovered that the crystalline lens was not the seat of vision, as had been supposed, but that its function, like that of other lenses, is the refraction of light. He observed that the image of objects is depicted upon the retina, and (with Scheiner) demonstrated that the expansion of the optic nerve in the retina is the essential part in the organ of sight. Obviously, also, interest in the anatomy of the eye, which these observations everywhere stimulated, was, in a great measure, aided by the researches of Newton on light and color.

About this time, too, Casserius and others studied the auditory apparatus and described the ossicles, the small muscles of the internal ear, and the semicircular canals; they even followed the acoustic nerve. By the researches of a number of French and Italian anatomists it was likewise established that the true seat of hearing lies within the internal ear, the external parts being merely of assistance in conducting sound.

Thomas Willis was one of the first to consider the brain as an assemblage of organs and to assign special functions to certain of its divisions; he thus became a pioneer in cerebral localization, although most of his conjectures were inaccurate or fanciful. The workings of the brain were also studied by Pinel and others, who observed that in certain conditions of mania or partial insanity some of the mental faculties—such as memory, judgment, imagination, or will—were abolished or suspended, while other faculties were preserved; hence it was inferred that each faculty must have its own seat. The views thus enunciated were carried to an absurd degree by Gall, and later by Spurzheim, who made an entirely new classification, believing the cranium to be molded in a reasonably exact manner upon the brain, and that, by inspection of the exterior, the character of a given individual could be read. They thus founded the pseudoscience denominated phrenology, which we now know has practically nothing to justify itself.

About the middle of the seventeenth century Glisson (a professor in the University of Oxford) recognized a property pertaining to all living tissue, which he termed irritability, and which he regarded as sufficient cause for all the phenomena of life; he enunciated certain views that, in times past, have had an important bearing upon the pathology of disease, but which were forgotten for sixty years or more until revamped by the Dutch anatomist, Goerter. It was the latter, with the great Haller, who, by a series of very ingenious experiments, elevated the suppositions of Glisson to the dignity of demonstrated facts. In 1747 the results of Haller's researches were published under the modest title of *First Lines in Physiology;* the author was, in fact, the great exponent of the doctrine of irritability in neurophysiology, and for this deserves to be remembered wherever the history of our art is spoken of. This theory of irritability was applied to pathology by Fabre, of Paris, who refuted the mechanical theory of Boerhaave on inflammation, proving that the latter proceeds not from obstruction of the capillaries, but from exaltation of their irritability. It was also applied in many ways by Bichat, who enjoyed a brief, though memorable, career. The theory of irritability, along with the truths established by John Hunter in his researches on the blood, made a very distinct advance in the physiological knowledge of the seventeenth century, and the researches of those who contributed so much to its advance are well worthy of study even at the present day. In this line of investigation should, perhaps, also be mentioned the names of Winslow, Albinus, the two Monroes, besides Vicq d'Azyr, and others already named.

I have so far discussed the development of theories and researches of individuals. During the earlier portion of the seventeenth century there happened something which gave to materia medica a remedy so valuable, and which attracted such wide-spread attention, that it deserves special mention. I refer to the discovery of that great febrifuge, Peruvian bark. Malarial fevers had been known as early as the time of Hippocrates, and were universally treated largely with purgatives, sometimes with venesections. There had been no notable improvement in the management of pyrexias of this class down to 1638, when the Countess of Cinchon, wife of the Viceroy of Peru, became a prey to a fever which nothing could remove. It is said a Spaniard learned from the natives the secret of the bark, and advised its employment, whereby the countess recovered her health. This is the generally received account, although it has been widely discredited, and Humboldt expresses decided doubts as to the source whence the first knowledge of the bark was derived. Be this as it may, however, it is certain that, in 1639, the countess and her physician, de Vega, imported into Spain a quantity of ground Peruvian bark, and distributed it to various persons, though it was not made an article of general commerce until ten years later, when it was exploited by the Jesuits, who had received a large supply; in Spain it was known as the "countess's powder," and in Italy as "Jesuit" or "cardinal" powder. Being very high-priced, it was soon so sophisticated as to be quite unreliable. Condamine, the botanist, having been sent to America for other purposes, determined the botanical position of the tree and described several species of cinchona, one of which is known by his name. To him is due the generic title bestowed in acknowledgment of the services rendered by the countess, who introduced the bark into Europe. Many vain attempts were made to determine the chemical composition of the powder, and it remained for two French chemists to isolate and separate its most important alkaloid. The first who wrote upon the therapy of cinchona was Barba, a Spanish physician, whose work was printed in Seville in 1642. After its introduction into England Peruvian bark fell into disrepute, owing to improper administration, whereby death was caused in certain instances; and it was this latter fact that instigated Sydenham to investigate it still more accurately. There has never been introduced into medicine any one drug which has proved itself so generally valuable and so widely effective as cinchona and its products.

As little progress had been made in obstetrics as in other branches of applied medicine or surgery. The custom of employing midwives was general, and these. for the most part, were ignorant and filthy old women, slaves of routine procedures that had obtained from time immemorial. Educated accoucheurs were called only in extraordinary cases; but with progress the prejudice which excluded educated physicians from the practice of midwifery gradually gave way, and there was opened for obstetrics a new era. In the beginning of the seventeenth century the initiative was taken by Louise Bourgeois, the *sage femme* of Marie de Medicis, who in 1626 published a collection of observations concerning sterility, abortion, fecundity, accouchement, and diseases of women and children generally; it embodied several distinctly new ideas. A little later (in 1668), Mauriceau, of Paris, chief accoucheur to the Hôtel-Dieu, published his treatise on diseases of pregnancy and childbirth, which was translated into all the languages of Europe and became a powerful agent for good, not alone that it represented an advance in knowledge, but it stimulated such rivals and successors as Devanter, Peu, Paul Portal, and Delamotte to further research. About this time the Chamberlains, an English family devoted to the practice of midwifery, invented an instrument to facilitate the extraction of the fœtal head when arrested, and one of them went to Paris, and, failing of success there, went on to Holland, where he sold his secret to two Dutch practitioners, who kept it only too faithfully. In 1721, Palfyn, a surgeon of Ghent, while seeking to fathom the device of the Chamberlains, conceived a *tire tête* (literally a head-drawer) composed of two steel spoons, and hastened to publish an account thereof,—a praiseworthy act, whereby he merits distinction as the inventor of the forceps. As modified by Smellie in England and Levret in France, the obstetrical forceps ranks among the most useful discoveries of modern surgery, and, although not in common use until about a century ago, it may be said that the invention has been the means of saving the lives of countless women and children.

Medical jurisprudence also seems to have had its beginning during this century. It had long been the practice to summon physicians to court in order to enlighten the judiciary in questions demanding particular knowledge in physics and medicine; indeed, the practice began under the first Christian emperors, and owes its origin to ecclesiastical authority. Charlemagne confirmed in this regard what Justinian was perhaps the first to ordain. The tribunal of Châtelet, according to Renouard, appears to have been the first which comprehended the great utility of consultation with expert physicians; an edict of Philip le Bel, in 1311, qualified Master John Potard with the title "Sworn Surgeon of Châtelet"; and the constitution promulgated by Charles V, in 1552, gave great importance to medical jurisprudence, as it treated in detail of infanticide, wounds, poisons, abortion, and other such crimes. Early in the seventeenth century Fidelis collated all that had been written on this subject, and thus published the first special treatise on legal medicine.

Some writers claim to have discovered traces of clinical teaching in the history of Arabian universities, but, as Renouard says, the presence of a few pupils during visitations and consultations no more constituted real clinical teaching than the practice adopted by some practitioners of ancient Rome of being ever surrounded by a group of spectators whom they dignified with the title of disciples. The first attempt at real clinical teaching appears to have been in the hospital of St. Francis, in Padua, in 1558, by Botoni and Oddi. About the beginning of the seventeenth century Otto de Heurne, of the University of Leyden, introduced bedside instruction, which was continued by le Böe, sometimes called Sylvius, with the result of drawing large crowds of students to Leyden from 1658 to 1672. Notwithstanding the success attained, the practice was neglected by the successors of Sylvius until renewed by Boerhaave, who, invested with several functions at the University of Leyden, also occupied the chair of medicine. So great was the renown of Boerhaave that, despite the poverty of the resources of the Leyden hospital, people came to consult him from the most distant countries, and he was a correspondent of several crowned heads, even of the Pope, although himself a Protestant. During his life and long afterward he exerted an immense influence in medicine, and while, perhaps, inferior in genius to some of his contemporaries, he had a wider reputation, and his doctrines prevailed longer. The great success of his clinics decided in favor of this method of teaching, and in 1715 the Pope established in Rome a similar institution, under the direction of the celebrated Lancisi. Soon Edinburgh, Vienna, Pavia, and other universities followed suit, the first clinical chair in Paris being held by Corvisart, and the first in Vienna by Van Swieten. After the demise of Boerhaave, the school of Leyden rapidly declined, while those of Edinburgh and Vienna became rivals for the first place. It is thus seen that after an interruption of more than two thousand years clinical teaching was revived and became more brilliant than ever before.

I now propose to recount the methods and deeds of some of those concerned in the development of systems, so called, and make mention of the most prominent medical men in national and historical order. This will not prevent going back to philosophical conclusions or reflections upon the philosophy of the history of medicine, when it may seem wise so to digress.

First, of the system of J. B. Van Helmont, which in its day was most highly regarded, and which seems to have been, in some measure, a rearrangement of the views of Paracelsus into a mystic and pietistic system based upon mechanical principles. Van Helmont was born in Brussels in 1578, and was so precocious that he entered the University of Louvain at an age which would have enabled him, had he so desired, to obtain the degree of Magister when only seventeen years old. He deemed the degree frivolous. He had studied mathematics, astronomy, philosophy, and astrology. Going now to the Jesuits, who at that time, even, taught music, he soon became dissatisfied, and turned to the study of stoical philosophy. Believing that the Capuchins (who were mere lascivious gluttons, and considered even washing unchristian) were the true stoics, he sought to join this order, but ere long abandoned them and resumed his studies in law, botany, and medicine. For the latter Van Helmont had at first little respect, since his studies in this line did not enable him to rid himself of the itch. He soon again lapsed to the monastics, and came to the conclusion that wisdom, like the grace of God, was obtainable only by fasting, supplication, and poverty; accordingly he practiced medicine among the poor as a labor of love (having received his degree of Doctor in 1599). During his travels he became familiar with the writings of Paracelsus, which he studied zealously. Finally he settled down in Vilvorde, where he practiced medicine and chemistry until his death (in 1644).

Like most "systems," that of Van Helmont is valued only as an expression of the spirit of the age, since it embodied largely the pantheism of Paracelsus, merely cloaked with a more religious or monkish dress. He held that the general cause of disease was the fall of man; though there also figured a subsidiary cause, which he denominated Archeus,—a faculty of appetite seated in the spleen or in the stomach; thus dropsy was a hindrance of renal excretion by the enraged Archeus. Demons, witches, and ghosts were included in Van Helmont's system as causes of disease. Indeed, the man seems to have been a second Paracelsus, lacking only in the dishonesty and bombast of the latter. He had no followers of any prominence, and the "system" soon lapsed into obscurity.

The Chemical, or Iatrochemical, System was originated by le Böe, commonly known as Sylvius (but who must not be confounded with the great anatomist of the same name). Le Böe was born in Hanau in 1614; studied in Paris, Leyden, and Basel; received his doctorate from the latter university at the age of twenty, and practiced in Switzerland with great success until 1660, when he accepted a professorship in Leyden; here he was distinguished for his eloquence, wealth, and sociability, as well as for the great number of pupils that were attracted by his clinical method of teaching. His system embraced a peculiar phantasy, being based upon the elements of chemistry, the new knowledge of the circulation, the latest physiological teachings, and the old doctrine of the spirituous or innate heat of the heart, which he claimed to have felt with his finger. He asserted his theories were founded upon experience, but the truth is, they were inaccurate deductions from experimental observations, many of which were wholly irrelevant. The majority of diseases, he taught, were produced by excess of acidity or alkalinity. For him, the three great fluids of the body were the saliva, the pancreatic fluid, and the bile, while health consisted in the undisturbed performance in the body of the process of fermentation; and the saliva was supposed to give rise to hectic fevers, because such manifest exacerbation after eating. Stereotyped theory and equally stereotyped therapeutics gained for him, for a short time, a large following, but later raised numerous opponents, who alleged that his system caused as many human lives as the whole thirty years' war. He died in 1672.

To the same iatrochemical school is generally assigned Thomas Willis, born in Oxford in 1622 (died in 1675), who rendered great service to anatomy, especially to anatomy of the nervous system, although his teaching was disfigured by certain unsupported theories. Like Van Helmont, he had been destined for theology, but turned his attention to medicine. Ultimately he became Professor of Philosophy in the University of Oxford. He first described the so-called circle of Willis, whence its name; also ascribed diseases, especially those of the blood, to fermentation, in which the vital spirits played the chief part. He accounted for hysteria, for instance, by the union of the spiritus with imperfectly purified blood.

CHAPTER VII.

AGE OF RENOVATION (*continued*).—*Iatromechanical School:* Santoro, 1561–1635. Borelli, 1608–1679. Sydenham, 1624–1689. Sir Thomas Browne, 1605–1682.—*Surgery:* Denis, † 1704. F. Collot, † 1706. Dionis, † 1718. Baulot (Frère Jacques), 1671–1714. Scultetus, 1595–1645. Rau, † 1719. Wiseman, 1625–1686. Cowper, 1666–1709. Sir C. Wren the Discoverer of Hypodermatic Medication. Anatomical Discoveries. General Condition of the Profession during the Seventeenth Century. The Eighteenth Century. Boerhaave, 1668–1738. Gaub, 1705–1780.—*Animism:* Stahl, 1660–1734.—*Mechanico-dynamic System:* Hoffmann, 1660–1742. Cullen, 1712–1790.—*Old Vienna School:* Van Swieten, 1700–1772. De Haën, 1704–1776.—*Vitalism:* Borden, 1732–1796. Erasmus Darwin, 1731–1802.

The physiology of the Iatromathematical, or Iatromechanical, or Iatrophysical School devoted chief consideration to the solid parts of the economy, whose form and function it strove to discover and demonstrate by the aid of exact methods,—that is, by calculation and physical apparatus. Thus, it explained digestion as mechanical trituration; secretions were referred to variation in resistance of parts in the vascular system; warmth was supposed to be due to friction of the blood-corpuscles; health consisted in the undisturbed performance of the physical and mechanical processes of the body. Diseases were explained inversely: the blood, under diseased conditions, was held to contain pointed and angular crystals, which irritated as they passed through the pores, or disturbed because they could not so pass.

The first to enunciate these views was Santoro, or Sanctorius, who flourished from 1561 to 1635, and was for a while professor at Padua. He taught how to investigate the pulse by an instrument of his own contrivance, and how to study the temperature by means of a species of thermometer, which was probably his own invention. (This instrument, by the way, was invented about this time; Drebbel [1572–1634] is regarded as the inventor of the air-thermometer, Galileo [1574–1642] of the spirit-thermometer, and Roemer [1644–1710] of the mercurial thermometer.) Santoro studied the phenomenon of transpiration, and constructed apparatus for bathing bed-ridden individuals; he found that in twenty-four hours the insensible transpiration through the skin amounted to 1¼ kilogrammes,—which result, compared with the results of the present day, determined by the most complete observations, is only twenty per cent. too high, and proves how accurately he investigated. The important rôle of the perspiration, which he pointed out, was made use of by the iatrochemists to vindicate their terrific sweat-cures.

Borelli (1608–1679), of Naples, is usually regarded, however, as the founder of the iatromechanical school. Of a quarrelsome disposition, he could not stay long in any one place, though he ultimately settled in Rome, where he joined the circle of *savants* who gathered round Christina, the daughter of Gustavus Adolphus, who had become a convert to Catholicism. Finally Borelli entered a monastery. His services related mainly to physiology, where, like Descartes, he followed purely mathematical principles; he explained the action of the muscles by the laws of the lever, calculated the mechanical work done by the heart, and correctly ascribed inspiration to muscular action. He was the opponent of iatrochemistry, and claimed there was no such thing as corruption of the blood. His pupils and followers—like Bellini (1643–1704), of Florence, who became professor in Pisa at the early age of nineteen, and Baglivi (1668–1707), a pupil of Malpighi, and a man of universal education—carried out and elaborated the first expressions of this author. Borelli was the author of the oft-quoted maxim: "He who diagnoses well cures well."

The iatromathematical system held ground for some time in Italy, and also found followers elsewhere. For instance, Dodart (1664–1707), of Paris, explained the voice on the mechanical principles enunciated by Borelli and by Quesnay (1694–1774), the first permanent secretary of the Academy of Surgery in Paris. In England this explanation was adopted by a number of followers, none of whom, however, was eminent enough to justify special mention here. In Germany it obtained a certain amount of favor, but seems not to have attracted any very eminent disciples.

The iatromechanical school ran a course not unprofitable to science, yet was unfruitful of real advance in the

FIG. 23.—THOMAS SYDENHAM.
(From a steel engraving of a painting by H. F. Rose.)

domain of practical medicine. The man of this particular age, who, more than any other, exerted an influence destined to be prolonged even to the present time, and probably much longer, who had a cool, clear, and unprejudiced spirit, and who sought the true value of medicine, and recompense for the same in the benefits which it brings to the sick, without scorning or neglecting its scientific side, was Thomas Sydenham, born at Winford Eagle in 1624, a student at Oxford in 1642, and recipient of a bachelor's degree of medicine in 1648.

The next fifteen years of his life we know practically nothing of, save that he spent some time in Montpellier pursuing his medical studies. In 1663 he became a member of the Royal College of Physicians, but did not take his degree of Doctor until 1676,—thirteen years before his death. His chief work—*Medical Observations*—is said to have been originally written in English, and translated into Latin; it first appeared in 1666,—the year when fire and plague devastated London. He died of gout in 1689, and was buried in Westminster Abbey. During the earliest years of the plague in London he fled, as was the general custom of that day.

His model was Hippocrates. In pathology he was a humoralist without being a theorist. He knew only one standard,—observation and experience. Sharing the opinions of his day, he laid but little weight upon anatomy and physiology; yet he recognized their value when employed in the production of hypotheses. He conceived of disease as active, operative,—a natural effort of the body to remove morbid material from the blood; if this effort is violent and speedy, we have to do, he says, with an acute disease, but if slow and difficult, the condition is chronic. Fever was supposed to result mostly from cold or from epidemic influences. As causes of disease, he considered unknown influences and changes of the atmosphere very important. In his special pathology "inflammation of the blood" played the chief rôle, and upon it were made to depend nearly all acute and some chronic diseases. He arrived at what he called the "healing power of Nature," for which he made great claims in his description and observation of epidemics: but he believed there always remained a good deal for the physician to do, and in treating syphilis he even gave mercury until two kilogrammes of saliva were discharged daily. As compared with the therapeutics of that day his were manifestly simple,—and yet he employed, for example, eighteen different herbs in one prescription, and that merely an ointment. The unreliability of the action of drugs induced him to rely upon specifics, as did Paracelsus, but he acknowledged only one such,—the then new discovery, cinchona,—not even allowing mercury such a position in the treatment of syphilis. Such drugs as he chose were mainly from the vegetable kingdom.

The great importance of Sydenham, and all his statements, so far as we are concerned, centres about his struggle for the elucidation of the healing power of Nature, and for simple observation and simpler treatment, as opposed to the overgrown luxuriance of previous systems and theories. He became the standard-bearer of his age in his return to Hippocrates's method and art of healing, which are founded on the nature of things and on the limits of human ability.

Sydenham was vehemently opposed by Richard Morton (1625-1648), of London, who, like Fernel, considered all diseases to be a poisoning of the vital spirits. Sydenham was also antagonized by Gideon Harvey, who ridiculed his medical contemporaries without stint, because most of them, for febrile disease, gave cathartics from the second day, and began treatment with emetics. With delightful satire Harvey divided the physicians of the day into six classes: the Ferrea, Asinaria, Jesuitica, Aquaria, Laniaria, and Stercoraria, according as their favorite systems of treatment were the administration of iron, asses' milk, cinchona, mineral water, venesection, or purgatives.

Sir Thomas Browne (1605-1682), who still enjoys a great reputation, was the author of the works entitled *Religio Medici* and *Inquiries into Vulgar and Common Errors*. The latter appeared in 1646, but does not seem to have protected its author from the worst error of his age,—viz., superstition,—since, in 1664, he swore that two condemned old women were actual witches.

Having considered the progress of medicine during the seventeenth century, it may be well to glance likewise at surgical progress. Among the Italians Santoro, already spoken of as the inventor of various instruments, should be mentioned; also Valsalva, who obtained a sound reputation as an operator, employed the ligature, and recommended a starvation plan for treating aneurism; Magati (1579–1647), who contended against the abuses of treating wounds by filling them with plasters, balsam, poultices, tents, etc., and of changing the dressing several times a day,—once in four days was better, he said; Severino (1580–1656), first a lawyer, then a professor at Naples, and later an eminent surgeon, a good anatomist, and a particular friend of the actual cautery; Marchetti (1589–1673), a bold, versatile operator of Padua; and Borri, of Milan (1625–1695), skilled as an operator and an oculist but better known because of his sad fate, since he died in the prison of the Inquisition, after a prison-life of twenty-five years, on account of too liberal religious views. There were also numerous other Italian surgeons who made a name, especially in plastic surgery, and particularly in that branch of it named rhinoplasty, by whose efforts one method of manufacturing a new nose came to be known as the " Italian method."

France, we must remember, was the home, during this century, of Richelieu, Mazarin, Louis XIV, Corneille, Racine, Molière, Fénelon, la Fontaine, Boileau, Bossuet, and many other men eminent in literature and science. During this century the French laid the foundation for that leadership in surgery which they maintained for nearly two centuries. Let us mention, among their surgeons, Morel, who invented the tourniquet at the siege of Besançon, in the year 1674. There was also Jean Baptiste Denis (who died in 1704), physician to Louis XIV, who performed the first transfusion of blood in man. (Transfusion of the blood of the young into the veins of the old, for the purposes of rejuvenation, was recommended by Libavius, in 1715, and Colle, of Padua, gave it new support by describing a method for its performance. In 1729 Boyle practiced transfusion on dogs. The London faculty sought the value of the operation after excessive hæmorrhage, and Edmund King, physician to Charles II, in 1665 practiced transfusion from vein to vein. But Denis was the first to carry out the operation with lamb's blood upon a patient sinking under excessive venesection,—an operation which was very much abused at this time.) It was in this century that the French family of lithotomists—the Collots—distinguished themselves in their special line. The last member of the family, François, died in 1706. Their specialty must have found, at that time, considerable more material than comes to the front to-day.

Among the general surgeons of France were de Marque (1618), who distinguished himself as a bandager; Bienaise, who invented the bistoury caché (1601–1631); de Launay (1649), monk and lithotomist; Goursaud, who survived his century, and who was the first to describe stercoral incarceration; Duverney, who demonstrated the growth and nutrition of the bones by periosteum; Lambert, who practiced injections in hydrocele; Andry, of Lyons, who wrote of orthopædic surgery and originated the name *orthopædics*; Pierre Dionis (who died in 1718), surgeon to the Empress Maria Theresa, famous in his art, and who first emphasized the effects of rickets upon the pelvis; and Boulot, better known as Beaulieu (1671–1714), who advanced himself from being a soldier and a day-laborer to become a physician, a famous lithotomist, and surgeon. He finally joined the Franciscan order, where he obtained the name of Frère Jacques, under which title he passed for the inventor of lateral lithotomy. Then there were Saviard (1656–1702), surgeon-in-chief in the Hôtel-Dieu, who, among other things, determined the seat of hernial strangulation to be often in the neck of the sac; and

Georges Mareschal (1658–1736), surgeon to Louis XIV, one of the founders of the Academy of Surgery, who has a record of eight lithotomies performed in half an hour, and who became famous for his services in improving the schools of surgery in France.

In this (the seventeenth) century, also, ophthalmology was much cultivated in France, although it was assigned to the despised surgeons. Those who won most renown in this line were Maitre Jean and Brisseau, who divide the honor of first recognizing the seat of true cataract. During this period, also, Duverney, Professor of Anatomy at Paris, was the first to systematically describe diseases of the ear in accordance with their anatomical seat.

In Spain scholarship sank more rapidly during this century than among any other people in history, due mainly to the loss of their political supremacy and their commerce to the Dutch and English, and to the utter failure, at home, of their efforts to introduce true unity of faith. In these efforts the industrious Moors were excluded, under Philip III. In art they maintained their standing,—attaining, in fact, in Murillo, the acme of their fame; but in other branches of industry they rapidly degenerated. Students of history will readily understand how little leisure the Spaniards had at this time to devote to the cultivation of science, including medicine and surgery. Of the two men who are mentioned during this century as Spanish surgeons, namely, Almeida and Ayala, we know practically nothing.

The Germans gained no such store of knowledge from their experience during the Thirty Years' War as did the French during their campaigns. The barber-surgeons, for the most part, still reigned supreme, and their guild contained some men of ability and independence of thought. The most notable man of the times was Fabricius Hildanus (1560–1634). Of him, however, I have already spoken as belonging rather to the previous century. He was the first

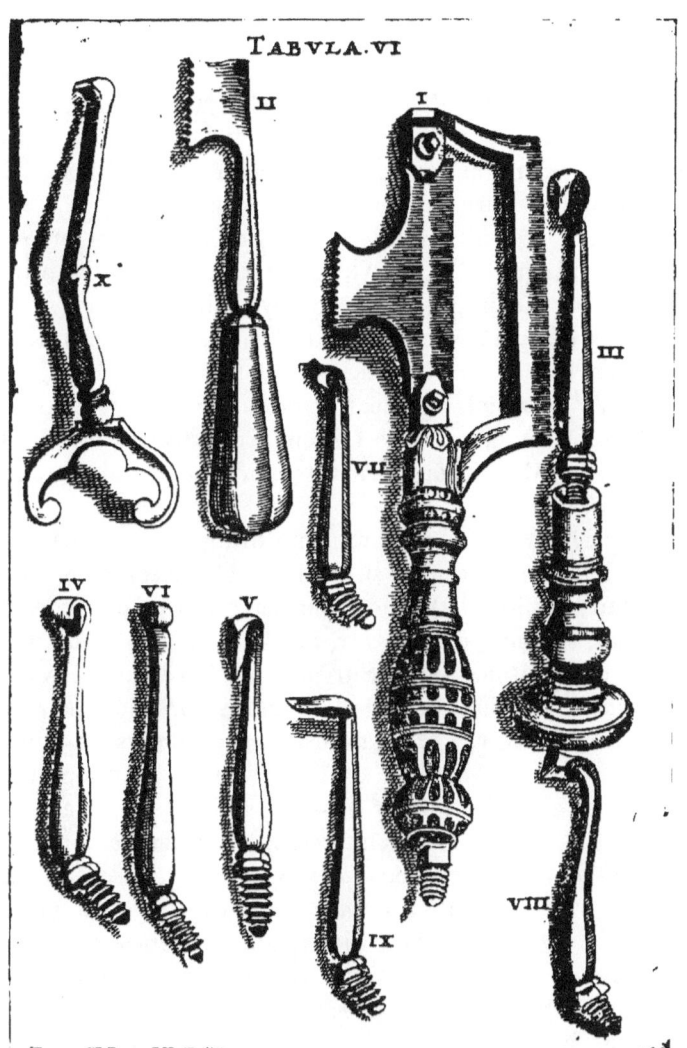

FIG. 29.—STRAIGHT SAWS AND DIVERS SCRAPING TOOLS, WHEREWITH THE SKULL, BEING ROTTEN OR HAVING A FISSURE, IS SCRAPED AWAY.

I and *II*, straight saws. *III* to *X*, various forms of scraping tools.

(From *The Chyrurgeon's Storehouse*, by Johannes Scultetus, a famous physician and surgeon of Ulme in Suevia. English translation published in London in 1674.)

learned German surgeon recognized and esteemed as such by his contemporaries. He was distinguished, also, as an oculist and aurist, and removed a particle of iron from the cornea by means of a magnet. A man of great operative genius, and a born surgeon, was Purmann (1648–1721), who greatly lamented the low condition of surgery in Germany, and regarded a knowledge of anatomy as the prime requisite for the surgeon; he employed the speculum in the diagnosis of syphilis, although it has been Ricord's boast that this was his own idea. Scultetus (1595–1645), of Ulm, was a famous surgical writer of this period, and a bandage of his devising is still in frequent use, and bears his name. Muralt, of Zürich, was also a capable surgeon (1655–1733).

The Dutch had but few men during this century who enjoyed any reputation as surgeons. The best among them was Rau (1658–1719), who, from being a poor boy, became a barber, traveled extensively, and was finally made Professor of Anatomy and Surgery in Leyden, where he introduced the innovation of teaching practical surgery upon the cadaver. He was especially famous as a lithotomist after the method of Frère Jacques, although he did not give instruction on this subject in his lectures.

By the way, it is an interesting fact that the clinical histories of many operations for stone during the seventeenth century were related in verse, and illustrated with plates. Harvey's vivisections were also related in verse.

Now, for the first time, do we begin to hear of English surgeons and English surgery. The most prominent, as well as almost the earliest, was Richard Wiseman (1595–1686), ordinary surgeon of James I, called sometimes the "Pride of England" and sometimes the "Paré of England,"—a bold, judicious operator, who took hold of every novelty and who accepted the ligature of Paré (always having the actual cautery at hand, in case the ligature should fail); he also amputated through sound

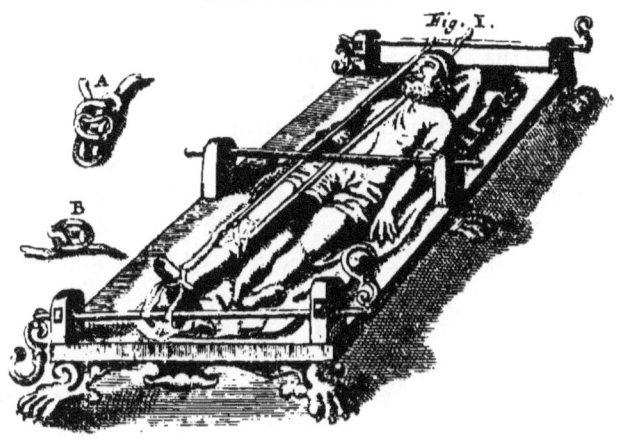
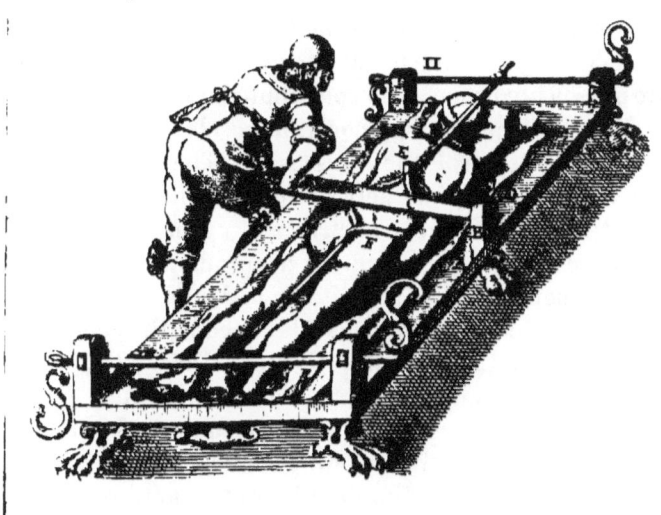

FIG. 25.—SURGICAL TREATMENT OF CERTAIN DISLOCATIONS.

I shows the reduction of the ankle-bone. *II* shows the extension and impulsion of the spina dorsi dislocated externally.

(From *The Chyrurgeon's Store-house*, by Johannes Scultetus, 1674.)

parts, favored operating for strangulated hernia, and employed the trephine zealously. The first recorded operation for external urethrotomy for the relief of stricture is mentioned in Wiseman's writings.

There were also William Cowper (1666–1709), a famous anatomist and surgeon; and Woolhouse, a famous, but ignorant, itinerant oculist. Sir Christopher Wren, architect of St. Paul's, was the first who devoted attention to injecting medicine into the veins,—a subject studied again much later and recently once more taken up. His example (in 1667) was followed by others, whose experiments demonstrated, as we know to-day, that the effects which follow the intravenous administration of drugs are the same as follow administration by the mouth.

Midwifery during the seventeenth century advanced even more rapidly than its mother-science surgery. The accouchement of women was intrusted in many cases to the care of educated men, who contributed not a little to the art. Anatomy and physiology contributed also their quota to a clearer knowledge of these diseases. The obstetric forceps were for so long a time kept secret that they were of small benefit at first to the obstetric art. Among the French who were especially prominent as promoters of midwifery must be mentioned Marguerite de la Marche, chief midwife of the Hôtel-Dieu; François Mauriceau, President of the College of St. Côme; Jules Clement Delamotte, who was also a skillful surgeon; and Portal, who first proposed version by one foot. Among the Germans a few midwives distinguished themselves as independent observers, most of all Justine Siegemundin, daughter of a minister, who devoted herself to midwifery with such success that she became court midwife; she recommended puncture of the membranes for the production of artificial delivery, and especially advocated bimanual version.

But, perhaps, the most significant advances were made in the direction of studies in anatomy, physiology, and pathology. The history of the circulation we have already taken up. After Harvey's time, and largely because of his researches, physiologists were divided into two parties with regard to the origin of life. These parties were known as *animists* and *animalculists*. It was largely by the later researches of Highmore (1613–1685) upon the anatomy of the testis and the epididymis, supplemented by those of Aubrey in Florence concerning the ovaries (which had been previously considered as female testicles), and the researches of Stenon concerning the muscular nature of the uterus, that a better knowledge of reproduction was established. De Graaf (1641–1673), a physician of Delft, Holland, pointed out the ovarian follicles, known to-day under his name, while Swammerdam (1637–1686) studied the comparative anatomy of the ovaries,—and was, by the way, the first to prove that the queen bee is a female. Needham, the London anatomist and physician, and Hoboken, of Utrecht, described more accurately the placenta and the coverings of the ovum.

Anatomical discoveries crowded along about this time. For instance, Wharton (1610–1673) discovered the submaxillary duct, named after him; Glisson (1647–1671) studied the liver and recognized its capsule, that still bears his name; Nuck injected the lymphatics with quicksilver, and studied the glands especially; Stenson discovered the excretory duct of the parotid, and Rivinius (his name being translated in German, Bachmann) found the sublingual duct; Peyer, Schafhausen, and Brunner, the latter a professor in Heidelberg, discovered the intestinal glands which bear their names; Wirsung, of Bavaria (who was assassinated in 1643 by another physician), discovered in the dissecting-room of Vesalius, at Padua, the excretory duct of the pancreas; Pacchioni found the bodies named after him in the dura mater; Havers, of London, discovered the synovial glands and the so-called Haversian canals; Cowper, already mentioned, discovered the small glands named after him, located in front of the prostate, and Bartholin yet other glands, in the labia, which bear his name; Meibom, professor in Helmstädt, discovered the small glands in the eyelids which are named after him. Besides these, many other discoveries might be recorded here, did time permit. One other, however, deserves to be mentioned, with which the name of Schneider (1614–1680) must always be honorably connected. He described the mucous membrane of the nose and demonstrated anatomically and clinically that not the brain, but this membrane, secretes the mucous discharge during fluxes from the nose. This overthrew at once and forever the ancient doctrine, which included so many and various "catarrhal" diseases. I might add also that the best and most complete description

---

Description of Fig. 26.—"Of the corruption of the bones of the arm and shin, even as far as the marrow; of the shin-bone broken with a wound and the bones sticking out and bound with swathe-bands brought circularly about; and of the cutting off of the end of the hand or foot. *I* represents the corruption of the bone and of the marrow of the shin-bone. *II* represents the shin-bone wholly corrupted and rotten. *III* represents the place where the corrupt bone was situated and was now pulled forth with the pincers. *V* is that shin-bone corrupted, which the patient laid up for a memorial. *VI* is the bone of the right arm corrupted. *VII* represents the bone of the arm totally corrupted and sharp, which was pulled away with the pullers, but by pieces, without any noise or pain. *VIII* shows the place where the corrupt bone of the arm lay, which was now pulled forth, which Nature filled up with a callous, so that the patient could perform country business without any impediment. The patient was a countryman of Pappatavia, whose arm a souldier broke in four places, without any wound, anno 1636. *IX* is a fracture of the shin-bone with a wound, and laying the bone naked. *X* is the bone of the shin with a wound, broken, with bones sticking forth, and bound with bands not crosswise, but circularly brought about and laid within the capsula as it ought to be. *XI* is a hand affected with a secret canker which is cut off in the sound part, namely at the end of the radius and cubit bone. *XII* is a hand that is sphacelated, which, being laid upon the block (*D*), is amputated in the sound ends of the radius and arm-bone with a chizel (*E*), contrary to Hildanus, with good success. *XIII* is a basin filled with oxycrat, in which swims a bladder, which, being wet, must be applied to the mutilated part. *XIV* are two swathe-bands wrapt together (*F* and *G*), whereof each hath two ends, to bind the arm, whereof the hand at the end is cut off. *XV* represents a foot that is sphacelated, which is taken off in the mortified part, near the sound part with a pair of pincers. The mortified part being removed, the rest of the putrefaction is consumed with red-hot irons until the patient feels the force of the fire. After this two plagets are anointed with Hildanus, his unguent Egyptiae, which are applied to the escar; lastly, long plaisters (*I*) being laid upon it, the foot mutilated is bound with a wet band (*K*) as far as the knee, as the hand is unto the middle of the arm. *XVI* are divers sorts of iron instruments and made red hot, both to consume the remainder of the putrefied part and are also fit to stop the flux of blood."

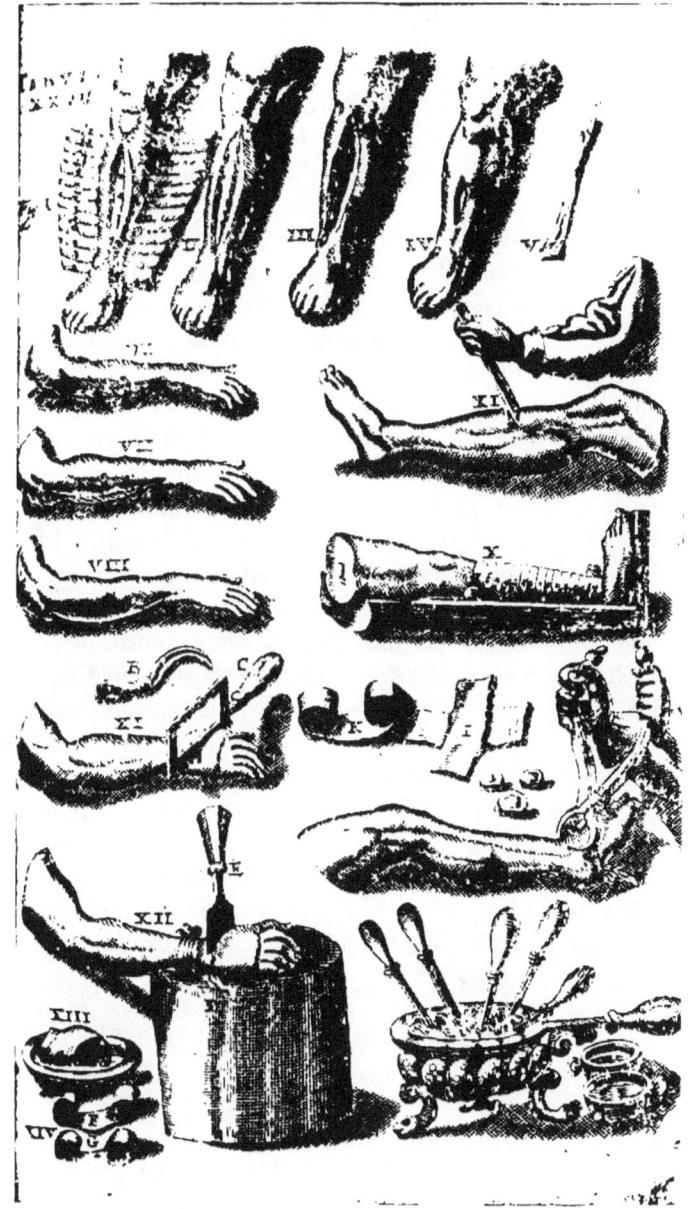

FIG. 26.—VARIOUS OPERATIONS ON THE ARMS AND LOWER LIMBS.
(From *The Chyrurgeon's Storehouse*, by Johannes Scultetus, 1674.)

of the entire central nervous system which had been given up to this time was furnished by Vieussens.

By the middle and latter portions of the seventeenth century most of the better physicians and surgeons had either assumed offices and positions in which they were supported by the State, or were settled in permanent residences, which was not the case with the mass of physicians in the sixteenth century. As a result the reputation of the entire profession began to improve, while the unlimited license and absolute freedom of practice prevailing during the Middle Ages were almost entirely done away with. By this time the clerical element had disappeared almost entirely from medical circles, or only dabbled in certain specialties. The Thirty Years' War was fatal to the supremacy of the clergy in matters of public health. Moreover, the increase of international intercourse favored the communication of medical knowledge.

The physicians of this period were more occupied with chemistry and physics than had ever been the case before. Nevertheless, this was also the special age of *alchemists* and of *impecuniosity*. According to one of the classifications of the time, the regular profession was supposed to include physicians, surgeons, barbers, regimental surgeons, lithotomists, bath-keepers, midwives, nurses, apothecaries, druggists, and even confectioners and grocers. Another list of impostors and quacks, equally official, was made to include old women, village priests, hermits, quacks,

---

Description of Fig. 27.—"*I* represents the breast affected with an ulcerated canker, the basis whereof is thrust through with two needles drawing after them a twisted flaxen thread. *II* shews how the chyrurgeon takes hold with his left hand, of the ends of the threads that were thrust through, and with his right hand he takes the knife and with that he cutteth the canker out by the roots. *III* shews a canker cut from the breast weighing six physical pounds. *IV* shews how the chyrurgeon, after the cutting off of a breast ulcerated, doth lightly cauterize the place with a red-hot iron at least to corroborate the parts. *V* is the instrument of Hieronm Fabritius ab Aquapendente wherewith a fistula of the thorax is perforated. *VI* is Sostratus, his band, which is most convenient where the breast is affected with any disease that requires binding. *VII* shews how Celsus cured the sticking forth of the navil by manual operation. *VIII* is a truss for the navil made of a double cotton linnen cloth."

THE ALCHEMISTS AND CHARLATANS. 187

uroscopists, Paracelsists, Jews, calf-doctors, executioners, crystallomancers (a class of people—chiefly Italian—who sought after crystals), mountebanks, vagrants, magicians,

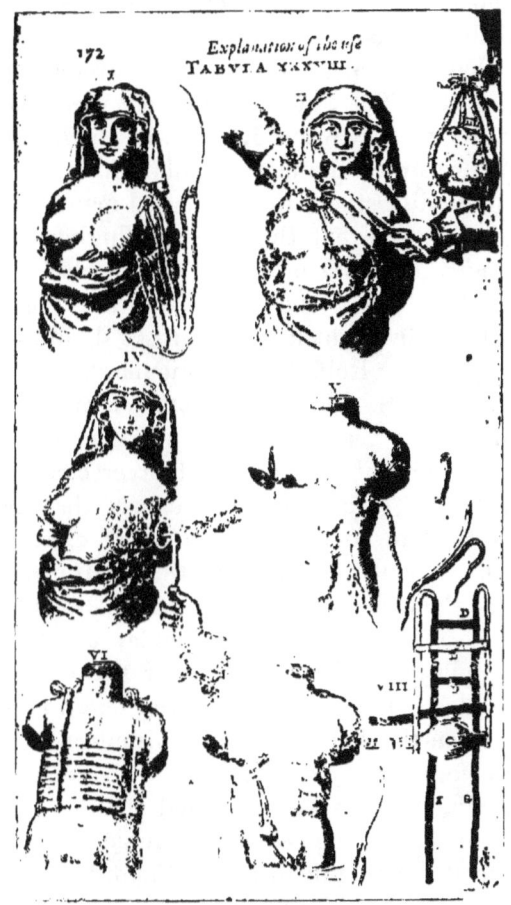

FIG. 27.—SURGICAL OPERATIONS ON THE BREAST, ETC.
(From *The Chyrurgeon's Storehouse*, by Johannes Scultetus, 1674.)

exorcists, monsters, rat-catchers, jugglers, and gypsies. Veterinary physicians were also at that time included in this class.

Anatomy was now studied more from human bodies, and was authorized by statute. This was especially the case in non-German institutions, to which for this reason students flocked in great numbers. In Dresden, so early as 1617, there was a dissecting-room in which stuffed birds, at that time a great rarity, and similar curiosities were preserved. The study of anatomy was at a low ebb in Germany; so that when Rolfink, in 1629, arranged at Jena, which was then the most popular German university, for two public dissections upon executed malefactors, it was considered such an event that the very highest authorities were present. But the peasantry took such fright at this occurrence that for a long time afterward they watched their cemeteries by night lest the corpses should be dug up and, as they said, "Rolfinked." Vienna did not possess a skeleton until 1658. Strassburg obtained one of a male in 1671, and several years later one of a female. In Edinburgh an anatomical theatre was first erected in 1697 in Surgeons' Hall. It is worthy of remark that anatomical plates, designed to be lifted off in layers, existed even at this period. About the middle of this century there arose a dispute at the bedside of the Margrave of Baden, between two learned professors and the regular court physician, whether a plaster to be applied over the patient's heart should be placed in the middle of the chest, according to Galen, or upon the left side. The dispute was settled by opening, before the eyes of the noble patient, a hog, by means of which it was demonstrated that, as a matter of fact, the heart of the hog lay on the left side. So convinced was his excellency that he dismissed the ordinary physician, who had held a contrary opinion as to the position of a nobleman's heart.

The general barbarity and immorality of this century were conspicuous, especially among the upper classes, and by its close had spread from France, became naturalized in both Germany and Italy, and extended even to the universities, their professors, and their students. The life of the latter during this period was more vulgar and rude than ever before, and almost more so than ever since. Pennalism—that is to say, barbarity toward junior students—became unbounded, so that outbreaks occurred even during lectures. At last the State authorities were compelled to interfere. Student outrages were very frequent and often fatal, and their outbursts were disgraceful in the extreme.

Only in France was instruction in surgery well regulated, for this was the only country which possessed a proper surgical college. Practical instruction was imparted to midwives—in Paris through a special institution, in Germany through the Midwives' Guild; the barbers, too, continued to receive instruction from their guilds; while instruction in pharmacy was given by the master-apothecaries, too often dogmatically and even farcically, serving as objects for the keen satire of Molière. The expenses of graduation were very great, and the ceremonies sometimes lasted two days.

In another way this same seventeenth century might be characterized as one of aggrandizement for physicians,—that is, as one during which their position was improved in the eyes of the public and better supported by the State. The physicians proper—the "*medici puri*"—were still persons of the profoundest gravity, with fur-trained robes, perukes, canes, and swords, when matters were prosperous, who for their lives would do nothing more than write prescriptions in formal style, everything else being considered beneath their dignity,—even as they affect in England to-day. They demanded to be called in every case, however, even though they knew nothing about it, claiming that only by means of their presence could things certainly go right. Nevertheless, in dangerous cases—for example, during the plague—they left the surgeons alone, while they looked upon the sick through the windows. In spite of this, however, they were generally esteemed and often sought for, as well in public as in private. Some of them were supplied with large libraries by their patrons or through their positions under the government, and most of them enjoyed moderate prosperity. Their pay was, for the most part, regulated in accordance with a definite tariff, while the State gradually cut down the doctor's honorarium to the pay of a day-laborer. During that century a certain physician to a countess in Munich received $25 as his annual stipend. For being present at a post-mortem and rendering an opinion thereon, each physician received $1.75. Surgeons who were zealous and eager were always highly esteemed; they were often better educated, in many respects, because of their extensive travels; but the social emancipation of the surgeons was not completed until the eighteenth century. About this time amputation of the arm was supposed to be worth 31 marks ($7.75); of the leg, 41 marks; or, if a patient died, half this price. Lithotomy cost 51 marks, or half of that if the patient died. For cataract operation on one eye the surgeon received 17 marks; for a like operation on both eyes, 25 marks.

We find in medicine, as in other branches of knowledge, that each succeeding century presents its added quota of imperishable facts, making it still more important than its predecessor. We may say that the fifteenth century had prepared the way for a reforming idealism which was the principal characteristic of the sixteenth; and that in the seventeenth century the realistic reaction against this same idealism showed itself in the church and the State by struggles against constituted authority, and in medical science by the domination of inductive philosophy. The idealism of the eighteenth century was not reformative and humanistic, but revolutionary and humanitarian. The unsettled character of the century's events may be charged, in some degree, to the American and French revolutions, with their interpretation (and their attempted attainment) of the so-called "rights of man." The masses were now supposed to be released, and philosophers created new doctrines, which had a greater influence upon the times than ever had philosophical doctrines before. Rousseau, for instance, aroused a revolution in politics and education, while skeptics and materialists alike strove for general enlightenment, which was sadly needed. Among the higher classes extravagance and immorality prevailed extensively, among the lower classes poverty and ignorance. In Germany the rulers even sold their subjects, as when Hesse-Cassel sold to the English seventeen hundred mercenary soldiers, and other States sold smaller numbers. A criminal code, published in 1769, contained seventeen copper-plate engravings, illustrating various methods of torture. A physician was always present when torture was inflicted, to see that the victim's sufferings were not greater than he could bear. This inhuman mode of eliciting testimony was last practiced in Europe in *1869*, in the Swiss Canton of Zug. Popular education was a myth, and the children of bondmen were not permitted to learn. No wonder the French revolution was hailed with joy along the Rhine, where it swept away at once and forever the petty rulers, abbots, and bishops, who were the "bloodsuckers" of the people. The numerous wars of the century had no great influence upon the development of medicine, except in the direction of surgery.

The eighteenth century was revolutionary also in the introduction of freedom of religious thought, so that clerical physicians disappeared entirely from the ranks, save a few who officiated as lithotomists, like Frère Côme, or as oculists, like Wrabetz, the latter of whom was even a professor in Prague.

This was the century, too, of Leibnitz and Kant, of Linnæus and Lavoisier, as well as of Bach, Haydn, Beethoven, and Goethe. During it the most conspicuous services in nearly all branches of learning were rendered by the

Germans, instead of by the Italians and English, as during the preceding century. In fact, Germany was then at the zenith of her glory, and supplied an impulse for all other nations.

The influence of philosophy and the natural sciences became also more and more marked. At the head of its philosophers must be placed Leibnitz (1646–1716), who, by his own writings and those of his pupils, created a philosophical school, whose influence is still every where felt. His doctrine was dualistic: Matter is created once for all, and has no further need of the Creator. As concerns the spiritual world, he assumed minute, indivisible, intelligent beings, called monads,—constituents of all bodies and all beings. In close relation with him stood Kant, while in England Locke and Hume became leaders of the opposed and materialistic school, declaring the brain to be an organ for the secretion of thought.

Among the universities founded during the eighteenth century were those of Breslau, 1702; Bonn, 1771; Stuttgart, 1781; Pesth, 1794; Göttingen, 1737; and Erlangen, 1743. Medicine was also cultivated in learned societies, which increased constantly in numbers. In 1744 Frederick the Great united two other societies into his Royal Academy. In Switzerland, in 1751, was founded an association of physicians and naturalists, while in France royal scientific societies were founded at Bordeaux, Montpellier, Lyons, and Dijon, and the Royal Medical Society of Paris lived from 1717 until 1788. In spite of all these opportunities for enlightenment, everything was not yet enlightened. Then de Haën defended the existence of demons, and Maerz, a well-known theological teacher, in 1760 devoted a book to witches and magic. That witches were burned publicly is a matter of history, even in America. So late as 1821 there was a statute regarding witches in Ireland, and they were burned in Mexico as recently as 1877. But these are flying pictures of the eighteenth century, which are meant only for the moment to illustrate the more serious topic, to which we must now address ourselves.

First of all, the *medical systems and theories of the century.* Many hundred years previously Galen had originated a method, which deserves, perhaps, the title of pure eclecticism. The first purely eclectic system similar to his originated with Boerhaave (1688–1738), perhaps the most famous physician of his or any other century. He was the

FIG. 28.—BOERHAAVE.
(From a steel engraving by Faustino Anderloni of a painting by G. Garavaglia.)

son of a clergyman near Leyden, Holland, and was one of thirteen children. Originally intended for the clerical profession, he had studied philosophy, history, logic, metaphysics, philology, mathematics, as well as theology, with great diligence. His education was, later, directed to the study of medicine, because of the statement that the purity of certain theological doctrines was endangered by him. So he studied chemistry and botany, and then anatomy and medicine, graduating in 1693. He practiced in Leyden with great success, and was offered a court position. In 1709 he was tendered the chair of Medicine and Botany, and in 1714 that of the Practice of Medicine; in 1718 he was also made Professor of Chemistry. In all of these positions he displayed the greatest capacity. He was a clinical teacher of rare talent, and soon acquired such reputation as to attract to Leyden students from all parts of the world in such numbers that no lecture-room in the university could contain them. He was the first to give separate lectures on the subject of ophthalmology, and employed the magnifying-glass in examining the eye. As a practitioner he was no less popular, and he left an estate valued at two million dollars. He was so famous that, when a Chinese official addressed a letter "To the Most Famous Physician in Europe," it reached him safely. He made no distinction in his patients, and compelled Peter the Great to wait a whole night for his turn to consult him. His most eminent pupils were: Haller, Van Swieten, de Haën, Gaub, and Cullen.

Boerhaave's influence and dignity, which were astonishing, even in a physician, were based no less upon his encyclopædic attainments than upon the benevolence and purity of his character. He was free from disputatiousness and vanity, although everywhere regarded as an oracle. His universal maxim was: "Simplicity is the seal of truth," although he never manifested this in his therapeutics. He employed the thermometer in the axilla in examining his cases, as did the iatrophysicists of the previous century.

His doctrines did not form a new system, but rather a composite of earlier systems. He stands also in the anomalous position of one who had the whole world at his feet, and yet contributed little or nothing which has been of essential importance. In fact, his peculiar views have been so universally given up that they are of only meagre historic interest. He looked upon disease as a condition in which bodily action or natural activities, being disturbed or unsettled, could take place only with difficulty; the reverse of this, of course, constituted good health. Fever he regarded as an effort of Nature to ward off death. Digestion was explained, like the circulation, upon mechanical principles. In his therapeutics, besides his efforts to sweeten the acid, to purify the stomach, to get rid of acridities, he made Hippocrates and Sydenham his models. His biographers say that his medicines were less effective than his personal appearance. He left many adherents, but no school of followers. It must be said, however, to his credit, that, while not the first to give clinical instruction, he permanently established a clinical method in teaching.

Gaub (1705–1780), professor in Leyden from 1731, was but little inferior to his master, Boerhaave, in fame as a teacher. He wrote the first complete work on the exclusive subject of general pathology. In general therapeutics he considered the healing power of Nature amply sufficient to remove sickness, but attributed this power sometimes to the soul and sometimes to the body.

There arose, naturally, strenuous opposition to the views and teachings of Boerhaave, and his principal opponent was Stahl (1660–1734), who was one of the most important systematists of any age, a profound thinker, and a pioneer chemist. He began lecturing in Jena at once upon his graduation, at the age of twenty-five, and moved through two or three different university positions until he came to Berlin at the age of fifty-six. He was a great pietist, of uncouth manners, faithful to his laboriously acquired convictions, and bitter and relentless against those who could not accept them. Indeed, he regarded his convictions as revelations from God. He looked upon the success of another as a personal injury to himself, and from being first a croaker he became finally a confirmed misanthrope, until he fell into actual melancholia. Pecuniary profit he had never sought, and its pursuit he scorned. His views were dynamico-organic, pietistic, and antagonistic. He regarded the soul as the supreme principle, life-giving and life-preserving, not to be confounded with the spirit; when hindered or obstructed in its operation, disease was present. The soul governed the organism chiefly by way of the circulation; consequently, plethora played an important rôle. To get rid of this plethora the soul employed either fever or convulsive movements; for example, in children plethora produces a pressure of blood to the head, and, by way of compensation, the soul provides a hæmorrhage from the nose. For reasons easily appreciated, he regarded bleeding piles as safety-valves of the utmost importance. Fever was a salutary effort of the soul to preserve the body; this was true even of intermittents, and, accordingly, he never gave cinchona. He scorned anatomy and physiology, saying, in one place, that medicine had profited as much by the knowledge of the bones in the ear as by a knowledge of snow which had fallen ten years previously. But Stahl was one of the most eminent chemists of the age, and did a great deal to liberate chemistry from the glamour of alchemy and the domination of pharmacy, and to transform it into an independent science.

Stahl's doctrine has been called *animism*, and was a reaction against the chemical and mechanical theories of the seventeenth century. He gained a considerable number of followers, the most notable of them among the French being Sauvages (1706-1767), the forerunner of Pinel and an opponent of pure mechanics, who animated the mechanical system of the body with Stahl's "soul." This was, par excellence, the age of artificial systems, and so Sauvages in his classification supplied a system which had ten classes of diseases, each of which had several orders, and some as many as two hundred and ninety-five genera, and two thousand four hundred species of disease!! Even Linnæus had three hundred and twenty-five genera of disease, while Cullen had only four classes with one hundred and forty-nine genera.

The mechanico-dynamic system was a sort of compromise or mixed system, which was held in high honor by the most eminent physicians and better minds of the last century, and has even been prized by Sprengel as the best of all. It was originated by Friedrich Hoffmann (1660–1742). Hoffmann's father was a physician, and he was himself born in Halle, whose university he attended. He acquired lasting reputation as an oculist, and was made Professor of Anatomy, Surgery, Medicine, Physics, and Chemistry at his *alma mater*. Our commonplace "Hoffmann's anodyne" is named after him. He was one of the most erudite professors of his day, more easily understood than Stahl, widely known for his fluent diction and amiable temper, and, accordingly, won great renown for his university. His good fortune as a practitioner was so great that even Boerhaave declared him his own equal. As a writer he was voluminous, one edition of his works comprising twenty-seven large volumes..

According to Hoffmann's views, life was simply mechanical movement, especially of the heart; death, the cessation of heart-action, putrefaction thereupon resulting. Health meant regularity of movements; disease, a disturbance of the same. He used the word "tonus" extensively. Ether he regarded as an important factor, producing and maintaining movements of the body, itself extremely volatile, corresponding largely to the "pneuma" of the ancients; it was, in fact, a motor principle and, at the same time, the perceptive soul. Ether was stored in the medulla, and circulated in a double way in the body; spasm was the consequence of too strong, atony of too feeble, influx of ether. Fever was a general spasm of the arteries and veins, having its cause in the spinal cord. Hoffmann's therapeutics were simple, and poor in drugs. These latter were intended to weaken, alter, or evacuate, and he was especially partial to the use of vinous remedies. The strong and toxic drugs he used but little.

William Cullen (1712–1790), a Scotchman, rose from the deepest poverty to the greatest celebrity. First a barber, he afterward became an apothecary, then a ship-surgeon, then a village practitioner, finally entering into partnership with William Hunter as a general practitioner. Both of these eminent men being in equally poor circumstances, they agreed to live in the same place and that, while one was studying, the other should take care of the practice. In this way Cullen was enabled to graduate in 1740. Six years later he taught chemistry in Glasgow, and in ten years more came to Edinburgh as Professor of Medicine. He continued very active and famous up to the time of his death, but died as he had been born,—in poverty. Among his numerous other charitable deeds, he supported the widow of Robert Burns and published the latter's poems.

Cullen was the father of modern Solidism,—a system based upon the solid parts of the body, the nerves being the chief agents. The life-giving element was, in his view, an undefined, dynamic something (different from Hoffmann's ether or Stahl's soul), which he called *nerve-force*, or *nerve-principle; animal force;* and *brain-energy*, and in it he included the spinal cord. His nerve-principle was supposed to produce spasms and atony, either actively or passively. The causes of disease, while of a debilitating character, were supposed to awaken reaction of the healing powers of Nature; fever was a reparative effort of Nature, even in its cold stage, the blood playing no part in it. He constructed a very arbitrary classification of fevers, as, in fact, he did of all diseases, his system of nosology being the secret of his reputation. His explanation of gout was famous. That disorder, he said, depended upon an atony in the digestive organs against which was periodically set up a reparative effort in form of a joint inflammation. In scrofula he had to assume, in contradiction to his nervous pathology, a peculiar acridity, and in putrid fever a putridity of the humors of the body. His therapeutics were simple and salutary, because of his renunciation of venesection, which was much abused in his day.

The most celebrated pupils and successors of Hoffmann were Gregory, of Edinburgh, Gardiner, and, in Germany, the famous Thaer (1752–1828), who finally abandoned the practice of medicine because it promised more than it could perform, and who became a "father of husbandry."

A composite of the doctrine of Hippocrates, Sydenham, and Boerhaave was represented in the so-called Old Vienna School, whose connection with the lives of Maria Theresa and Joseph II deserves, at least, mention. Its founder was Baron Van Swieten (1700–1772). of Leyden, a descendant of a noble Jansenist family of the Netherlands, who graduated under Boerhaave after having studied at Louvain. After the death of his patron he was called to the assistance of the Archduchess Maria Anna, of Austria, who was suffering from an abortion, and gave such satisfaction that she recommended him to her sister, Maria Theresa, who up to this time had remained sterile. To her and to her husband he gave advice which resulted in sixteen successive pregnancies, and then, as the result of his success, came to Vienna in 1745 as President of the General Medical Department of Austria. He was also made censor, in which position he incurred the enmity especially of the Jesuits and of Voltaire, whom he robbed of their influence. He was made baron, and became, next to Kaunitz, the most influential counselor of the empress. His chief care was dedicated to the elevation of medical affairs in Austria, and especially to the improvement of the medical faculty. He had just seen success crown his efforts when he died of senile gangrene, with the reputation of being a great physician and benefactor of the poor. One of the greatest of his services was improving the treatment of syphilis, in which he, after the example of Paracelsus, recommended the internal use of corrosive sublimate.

More eminent as a physician than for personal character was de Haën (1704–1776), of The Hague,—a pupil of Boerhaave. At the suggestion of Van Swieten, he was called, in 1754, to Vienna as president of the clinic of the city hospital, which at that time afforded accommodation for only twelve patients. He was the real founder of the so-called Old Vienna School, whose merit, in contrast to the so-called new school, is to be sought in practical and diagnostic services. As de Haën quarreled with every one, he also did with Stoerck (1749–1803), the successor of Van Swieten in the direction of the Austrian Medical Department, and with Stoll (1742–1787),—a clinical teacher who was especially famous as an epidemiologist.

Stoll lectured with great popularity until 1784, upon the completion of the Allgemeines Krankenhaus, when he fell into the background and was badly treated. He was the subject of numerous intrigues by his enemies, and had a wife who embittered his life, and who even had him buried in the dress of a Jesuit in order to injure his reputation after his death. To his credit be it said that, changing his views of the constituents of disease later in life and his original therapeutics becoming no longer of use to him, he abandoned them entirely. Nevertheless his therapeutic system flourished for a long time after him.

There were in vogue during this period numerous other doctrines, some of which were too puerile or insubstantial to gain any foothold at all; others exerted a certain amount of influence during the life-time of their originators or for a generation afterward. With many of these I do not care in any way to deal. A few others, I think, ought to be at least mentioned in such a history as I am endeavoring to present.

There was another Hoffmann—Christopher Ludwig

Hoffmann (1721–1807), of Westphalia, who devised a so-called humoral theory in which the "acridities" of Boerhaave were mingled with the "putridities" of the pneumatists and the "irritability" of Glisson. His treatment and remedies for diseases were supposed to be antiseptic, as was very proper when dealing with putridities.

The theory known as the "Doctrine of Infarctus" had its origin with Kampf, who died in 1753. By infarctus Kampf understood impacted fæces, which he thought originated in the humors of the body, portal vessels, and intestines; he recognized two kinds,—the black bilious and the mucous. From this theory a wide-spread clyster fashion developed, and lords and ladies vied with each other in belaboring their infarcti and in administering enemas. As Baas says: "We cannot deny to the author of this doctrine at least an extensive knowledge of human nature. He supplied a universal remedial procedure, and gratified the apothecaries with the bulkiness of the herbs required for its practice."

Quite antagonistic to the views of the Vienna School were those of the School of Montpellier, inaugurated by Bordeu (1732–1796), and generally known as *vitalism*. Bordeu died in the enjoyment of great reputation, but at variance with all his colleagues. He maintained the existence of a general life of the body,—a composite life,—resulting from the harmonious working of the individual lives and powers of all the organs, which were supposed to be associated with each other, but each for its own definite function; the most important organs—the stomach, heart, and brain—being called "the tripod of life." In pathology he laid great weight upon crises, which were supposed to proceed from the glands.

The most important representative of vitalism was Barthez (1734–1806), of Montpellier,—a man of great gifts and eager for knowledge. He recognized a vital principle as the cause of the phenomena of life, but acknowledged that its nature was unknown, although he endowed it with motion and sensibility different from a thinking mind. Plants were supposed to possess it likewise. Disease, he believed, was the result of an affection of this vital principle. Every disease was divisible into certain disease-elements, viewed as parts of the whole, and these were again divisible into secondary elements. He explained putrid fevers as specific vital diseases,—in which view, of course, he embodied humoral ideas.

In Germany, at about this time, a similar doctrine obtained,—a doctrine of vital forces,—which the versatile Reil (1759–1813) elaborated into a system.

Meantime, in England, a doctrine was elaborated by Erasmus Darwin (1731–1802) which partook, in a certain degree, of the doctrines of Stahl, Hoffmann, Haller, Brown, and Bordeu. Erasmus Darwin distinguished himself, not only as a physician, but as a poet, philosopher, and physiologist. He was a friend of James Watt. Of his life it is said that by his practice and very fortunate marriages he became wealthy, ate much, and drank nothing but water. His chief work—entitled *Zoönomia, or the Laws of Organic Life*—was published in 1784, and is well worthy of perusal to-day. He recognized two fundamental substances —spirit and matter. But it is not so much for his doctrine as for his researches into animal and plant physiology, and, reflexly, because of his more celebrated descendant of the same name, that we owe him most gratitude.

CHAPTER VIII.

AGE OF RENOVATION (*continued*).—*Animal Magnetism:* Mesmer, 1754-1815. Braid.—*Brunonianism:* John Brown, 1735-1788.—*Realism:* Pinel, 1745-1826. Bichat, 1771-1802. Avenbrugger, 1722-1809. Werlhof, 1699-1767. Frank, 1725-1801.—*Surgery:* Petit, 1674-1750. Desault, 1744-1795. Scarpa, 1772-1832. Gimbernat, † 1790. Heister, 1683-1758. Von Siebold, 1736-1807. Richter, 1742-1812. Cheselden, 1688-1752. Monro (1st), 1697-1767. Pott, 1749-1787. John Hunter, 1728-1793. B. Bell, † 1806; J. Bell, † 1820; C. Bell, † 1842. Smellie, 1680. Denman, 1753-1815.—*Revival of Experimental Study:* Haller, 1708-1777. Winslöw, 1669-1760. Portal, 1742-1832. Vicq d'Azyr, 1718-1794. Morgagni, 1682-1772.—*Inoculation against Small-pox:* Lady Montagu, † 1762. Edward Jenner, 1749-1823.

DURING the eighteenth century also arose the illusory doctrine of Animal Magnetism, which obtained among all classes a following that can be accounted for only by the attractiveness of the marvelous and unexplained. Frank Mesmer, born near Lake Constance, in 1754, was early a victim of romantic yearnings, and his graduating thesis, delivered in Vienna, dealt with the influence of the planets upon man and the use of the magnet. After traveling extensively he erected a private institution, where he treated blind girls, fidgety old maids, and simpletons, until his deceptive methods were unmasked by a commission appointed by the Empress Maria Theresa, and he was compelled to leave Vienna in twenty-four hours. This martyrdom recommended him in Paris, where the so-called Mesmerism speedily became fashionable. He finally undertook instructions in magnetizing, at the rate of 100 louis a head, and founded the "Order of Harmony." His so-called *baquets* were tubs with magnetic ducts, partially filled with soft water and all kinds of ingredients, and armed with iron conductors, with which his pupils, joining hands, placed themselves in contact. At these *séances* Mesmer appeared in lilac-colored clothes and professed to reinforce the action of the tubs by looks, gestures, playing upon the harmonica, and touching the subjects with wand or fingers. "If any one, particularly a lady, had a crisis at this time, she was borne to the 'crisis-chamber' by Mesmer himself, where he treated her alone, as only when alone, he claimed, could he attain success." He speedily became wealthy; managed to deceive even the Queen of France; and, when he threatened to deprive the country of his presence, 20,000 francs were offered him to instruct others in his art. This offer, however, the wily charlatan declined. In 1785 some fool penned an article extolling him as a worker of miracles; this stimulated the authorities to organize a committee of investigation, the adverse decision of which, along with some contributory evidence, made Paris too warm for him. After the revolution he returned, but his day had passed, and he figures no more in medical history. He has had many imitators, and the mesmeric craze, at times, has infested different portions of the civilized globe; even some who were eminent in science have fallen into the snares of so-called Mesmerism,—notably Olbers, the discoverer of a number of asteroids. Mystic medical doctrines, founded upon Mesmer's views, still continue in certain circles, though the majority have long since succumbed to the advances of scientific psychology. In this connection it is proper to speak of the revived interest in "animal magnetism" due to the researches of Dr. James Braid, of Manchester, England. This gentleman, in 1842, published a work which pretty thoroughly exposed the fallacies of the doctrine of Mesmer, and expounded many of the truths that were entangled therein. He was among the first, perhaps, to employ the phrase "animal magnetism," and was the author of the term "hypnotism," though in his day the popular title was *Braidism*.

During the middle of the eighteenth century arose a doctrine that, in its novelty, ease of practical application, and apparent consistency (through the ingenious employment of certain vital phenomena), secured such a hold that its influence continued even into the present century. This was the "Brunonian doctrine," promulgated and upheld by the great foe and rival of Cullen,—Doctor John Brown. In youth very precocious, though of most humble birth, Doctor Brown had mastered the Latin language at the early age of seven years, and three years later essayed to learn a trade. At the age of twenty he left his native village of Dunse for Edinburgh, seeking employment as a tutor and intending to study theology. Poverty soon com-

FIG. 29.—JOHN BROWN, M.D.
(From a steel engraving by J. Caldwell of a miniature painted by Donaldson.)

pelled him, however, to take a rural school, but he returned a few years later (in 1759) to the Scottish Athens and began the study of medicine, supporting himself meantime by rendering theses into Latin and by teaching, translating, and quizzing. Finally, he attracted the attention of Cullen, to whom he became useful through his knowledge of the classics; but, ultimately, a foolish quarrel made bitter enemies of the former friends. In 1770, in private lectures, Brown began to advance the theory to which he had been led by one of his own attacks of gout that disappeared under the use of stimulants, the disease having previously always been aggravated by the treatment prescribed and that was held to be orthodox,—viz., antiphlogistic. He had now become somewhat dissolute, and the students he gathered about him were of very much the same character; but they formed the nidus of a great following opposed to Cullen, and quarreled on all occasions with the adherents of the latter. Finally, Doctor Brown removed to London, where fortune seemed to smile upon him, as he gained rapidly in reputation and practice; indeed, he barely missed a call to Berlin and another to Padua as a teacher, the scale being turned against him by his dissolute habits. Though possessed of the highest mental gifts, Brown was unfortunate in lack of mental stamina. He taught that life is not a natural condition, but an artificial and necessary result of constant irritations; all living beings, therefore, tend toward death. Health is an intermediate grade of excitement; diseases, which are either sthenic or asthenic, represent either too high or too low a grade of excitement. It has been said that Brown's teachings slaughtered more human beings than the French Revolution and the wars of Napoleon combined. In England this system found no important followers, but in America Benjamin Rush, of Philadelphia (1745–1815), distinguished himself as an adherent. In Spain and France it found little place; but in Italy, and later in Germany, it secured a numerous and important following, which numbered, among others, Scarpa, Massini, and Girtanner.

Another system which attained influential development, extending even into the present century, was the so-called *Realism*, originated by Pinel (1745–1836). Born in poverty, and designed for the Roman Catholic Church, Pinel did not turn his attention to medicine until his thirtieth year, but on completing his studies he rapidly rose to positions of importance. Led to the investigation of mental diseases by the fate of one of his particular friends, who had become insane, escaped into the forest, and was there devoured by wolves, Pinel speedily developed a great interest in this class of sufferers. The lot of the insane at this time was most pitiable: they were imprisoned, chained, and treated worse than wild beasts. In his efforts to im-

FIG. 30.—PH. PINEL.
(From an old lithograph of the eighteenth century.)

prove their lot, Pinel acquired the title of conservative and aristocrat, either of which was almost equivalent to a death-sentence. Unterrified, however, he appeared before the Paris Council and urged the adoption of reformatory measures, replying to the challenges of skeptical and self-regardful opponents by liberating a number of insane patients who were in his charge. The courage thus exhibited receives appreciation in our time, if never before. Not the least of Pinel's services was the substitution of analytical for synthetical methods; he also sought to determine disease by a diagnosis carefully constructed from symptoms, but unfortunately he made pathology and anatomy subordinate factors. He was a pupil of Barthez, but he placed his preceptor's vitalism far in the background.

François Bichat, born in 1771, earned high rank both as a clinician and an anatomist. His education was begun in Nantes, but he studied surgery and anatomy in Lyons and Montpellier, subsequently going to Paris, where he became a member of Desault's family. After the death of his patron he lectured on surgery, and from 1797 on anatomy. Possessed of a feverish scientific activity, he became a member of the Société d'Emulation. Death overtook him in 1802 as the sequel of consumption and an injury received through a fall. He was the most capable physician of France in his time, and, brief as was his span of life, he was author of nine important volumes, the chief of which were a *Treatise on Membranes* and works on general and pathological anatomy. From the latter a new tendency in study took origin. He it was who gave utterance to the aphorism: "Take away some fevers and nervous troubles, and all else falls to the kingdom of pathological anatomy." As an evidence of his energy, it is related that he in one winter examined seven hundred bodies. He taught how to discriminate between disease processes, and notably subdivided peripneumonia into pleurisy, pneumonia, and bronchitis, these having been previously confounded. He once remarked: "You may observe disease of the heart, lungs, abdominal viscera, etc., night and morning by the sick-bed for twenty years, yet the whole furnishes merely a jumble of phenomena which unite in nothing complete; but if you open a few bodies, you will see the obscurity speedily give way,—a result never accomplished by observation if we do not know the seat of the disease." To Bichat is also due our modern recognition of cellular, osseous, fibrous, and other tissues, as such, wherever they appear throughout the body. He differentiated, without the aid of the microscope, twenty-one different tissues as simple and similar elements of the body, enumerating them as one does the chemical elements; he described the stomach as composed of mucous, serous, and muscular layers; overthrew the speculative tendency of medicine, and placed facts in the front rank; and so

FIG. 31.—MARIE FRANCOIS XAVIER BICHAT, M.D.
(From a steel engraving by H. Cooke of a painting by Vigneron.)

conspicuous were his services that he has been termed the "Napoleon of Medicine." He supplemented the influence of Pinel upon the side of pathological anatomy; called sensibility and contractility vital properties, whose alterations constitute disease, claiming, however, that the vital properties of individual tissues differed among themselves. His life and works are revelations to young men, and show what can be accomplished at a very early age by sufficiently active and harmoniously developed brains.

In reviewing the theories and lives of those mentioned as medical luminaries of the eighteenth century, one experiences a feeling of mingled respect and disappointment—respect for the devoted way in which they worked and sought for the truth, and disappointment at so much waste of intellectual power and labor. The lesson is also taught, and should be impressed, that in all so-called new systems old principles for the most part reappear, and that the labors of the past are rarely so deliberately consulted as to guard against repetition and revamping of theories that had long before been proved futile.

Let me now mention a few other of the physicians of the last century who have left more or less of an impress upon their successors and upon our science. One man, in particular, historians are wont to remember with the honor that was denied him by his colleagues and contemporaries. I refer to Leopold Avenbrugger, who was born in Graz in 1722, and who, after pursuing his philosophical and professional studies in his native city, obtained, at the age of twenty-nine, charge of a Spanish military hospital; while thus employed he invented the art of percussion as applied to diagnosis. This he gave the test of experience during seven long years before making it known to the profession, and even then it was not appreciated, but remained practically unnoticed until after his death, which occurred in 1809. He did receive a patent of nobility from the Emperor Joseph II, but this hardly compensated him for the contumely heaped upon him by his colleagues. Paulus Ægineta employed sounds and specula; Santoro used the balance, counted the pulse, and resorted to the use of the thermometer; Boerhaave employed the thermometer and the simple lens; Floyer, and after him Haller, utilized the watch in marking seconds; a Salernian practitioner utilized auscultation and percussion in tympanites and ascites; but the diagnosis of diseases of the great viscera by percussion was never known before Avenbrugger. His booklet of twenty-two pages, unsalable in his time, is to-day held worth far more than its weight in gold. His famous colleague, de Haën, wrote fifteen volumes without a word on percussion; Van Swieten did it no greater justice; in his great treatise on the *History of Medicine*, Sprengel barely alludes to it; yet the contents of Avenbrugger's booklet were of more practical value than all that these other men ever wrote, or all the results of the vast and bloody campaigns during which it slept. In 1808 this volume was rescued from oblivion by Corvisart, who translated it into French and proclaimed its undying value.

During the earlier part of this century lived Werlhof, of Helmstädt (1699-1767), a far-famed observer, author, and practitioner, who declined a professorship, and especially distinguished himself as a writer of German poetry. Though possessed of an exceptional knowledge of modern tongues, he wrote only in Latin,—the scientific language of the day. In 1734 he was appointed physician to King George II, in which position he attained world-wide fame, while indefatigable in his efforts to elevate science. He first described the disease known by his name,—*morbus maculosus Werlhofii*,—and struggled hard to establish in Germany the use of cinchona.

From 1740 to 1802 flourished Wichman, of Hanover, highly esteemed as a writer and practitioner. He is especially known for his pleas in favor of more scientific diagnoses, and his demonstration of how to make them. The rôle of the itch-mite in the transmission of scabies he demonstrated upon himself; to be sure, Bonomo, a hundred years before, had called attention thereto, but with little avail.

Another eminent Hanoverian was the fickle, stubborn, and misanthropic Zimmerman, born in 1728, in Berne, upon whom misfortune and disease played many shabby tricks. He was, however, a man of ingenious endowments, and merits especial regard, because he sought to free medical science from the charge of being a secret art.

Another of the prodigies of medical history was J. P. Frank, born (1725) in the Bavarian Palatinate, of pauper parents, and, while an infant, abandoned by a cruel father. His early life was passed in a religious school; at twenty-five he became a court and garrison physician, and later a professor in Göttingen; finally he went to Vienna, where he died in 1801. He was greatly beloved by his pupils, and Walther, the famous surgeon, said of him: "No one ever made so elevating and permanent an impression on me." He published an extensive work on forensic medicine and sanitation,—wherein he took up the hygiene of the individual, of the family, and of the school,—which constituted an effort far ahead of anything of the kind previously known. He is also memorable for efforts toward increasing the population, for the Thirty Years' War had depopulated extensive districts—to such a degree, indeed, that in 1750 bigamy was legalized in Nuremberg and many other towns. Frank was distinguished for a keen and even caustic humor, whose subject was not infrequently himself.

From 1707 to 1782 there lived in England one Sir John Pringle, chief of the Army Medical Department, known to this day as an author upon military hygiene. John Huxham (1794–1868) advanced our knowledge of putrid dissolution of the blood. John Howard (1766–1790) rendered eminent service in prison reform. Heberden (1710–1801) was the first to describe varicella, and also angina pectoris—which was long known as Heberden's asthma. John Fothergil (1712–1780), a Quaker, acquired fame by his observations on chronic angina, neuralgia, and hydrocephalus; was likewise a benefactor of the poor, regarding them as "bridges to the pockets of the rich"; indeed, a large part of what he gained from the latter class he bestowed in charity, and at his death left £200,000 for the same purpose. Radcliffe (1750-1814) was an eminent, witty, successful practitioner of London, who was wont to declare that, as a young practitioner, he possessed twenty remedies for every disease, but at the close of his career had found twenty diseases for which he had not one remedy. Richard Mead (1673-1754) was a prolific writer, and the author of the first quarantine regulations adopted in England. Contemporary with Mead was Lettsom,—the busiest, most philanthropic, and most successful physician of his day,—whose practice, although a large part of it was gratuitous, brought him sixty thousand dollars a year, and who gave away immense sums for charitable purposes; also, Thomas Dover, who invented the sedative known by his name and who died in 1741. Akenside, physician and poet (1721-1770), wrote on dysentery. Baillie, of Edinburgh, was the first to accurately describe the morbid anatomy of gastric ulcer.

Among the French surgeons must be mentioned la Peyronie, of Montpellier, born in 1668, who ultimately became director of the Academy of Surgery and surgeon to the king. His wealth was employed for the elevation of the craft, and he founded no less than ten different surgical professorships at his own expense. In 1743 he effected the separation of the surgeons from the barbers. He died in 1747, dedicating his estate to the purpose for which he had lived. The most famous of the earlier surgeons of this century was J. L. Petit (1674-1750), inventor of the screw tourniquet, and who was called to treat Augustus the Strong, of Poland; indeed, several other crowned heads became his patients. Garengeot (1688-1759), a professor in the College of St. Côme, published a work on operative surgery. Morand (1697-1773) and le Dran were distinguished surgeons of Paris, the former especially noted for the number of times he performed paracentesis. Famous lithotomists were le Cat and Frère Côme,—whose real name was Baseilhac, and who operated by means of the lithotome caché. Astruc (1685–1766) was a syphilographer of extensive attainments; Quesnay (1694–1774), an eminent and undaunted surgeon of Louis XV, who wrote on the history and progress of surgery in France; Brasdor (1721–1776) was best known for his method of distal ligation in aneurism; Sabatier (1732–1811) wrote a famous treatise on operations, in which he recommended resection of the head of the humerus.

One of the most celebrated surgeons was P. J. Desault (1744–1795), the son of a poor farmer, originally designed for the priesthood, but who, after obtaining a thorough mathematical education, began the study of surgery with an ignorant master of his native town. Subsequently he went to Paris, and here supported himself by teaching, gradually rising, step by step, until, without collegiate education, he became professor and chief-surgeon at the Hôtel-Dieu, where he established the first surgical clinic. He opposed violently the prevalent abuse of the trephine, and was also a champion of healing by first intention. A trusted friend of Desault was Chopart, well known because of the amputation of the foot that bears his name. Another well-known surgeon, likewise a friend of Desault, was Doublet; and it is somewhat remarkable that Desault, Chopart, and Doublet suffered persecution and perhaps martyrdom in connection with the supposed death of the Dauphin of France,—properly Louis XVII,—in 1795. There is evidence that the child who died in the temple was not the dauphin, but a substitute, and these three surgeons, who examined the corpse, had the hardihood to express their doubts. The same day that Desault reported upon the evidence he was invited to dinner by some members of the Convention, was taken ill at the table, and died almost immediately after his return home. A few days later Chopart and Doublet died, also under mysterious circumstances.

Daviel (1796–1862) is remembered among French surgeons chiefly for extraction of the lens as an independent method of treating cataract; Tenon (1724–1816), for his writings on the anatomy and diseases of the eye; and Anel for originating the operation for aneurism, mistakenly attributed to Hunter. There were also many others, of lesser note, who distinguished themselves through special services to surgery or some of its branches.

Among the Italians of this century may be mentioned Scarpa (1772–1832), of Motta, professor successively in Modena and Pavia, and who advanced our knowledge of hernia, diseases of the eyes, aneurism, and general anatomy.

The most famous Spanish surgeon was Gimbernat, of Madrid (1742–1790), for a time professor in Barcelona, who also became distinguished through anatomical researches.

German surgeons did not rank high during the earlier half of the last century, owing to the contempt engendered by the church for this branch of the medical art. The fashion of imitating the French, however, led to some surgical development. The first German surgeon of scientific education was Heister (1683–1758), of Frankfort-on-the-Main, who, unable to obtain honorable employment in the military service of his own country, entered that of Holland, where he remained until the experience of his own nation had brought about a healthy reaction. In 1720 he came to Helmstädt, where he developed great activity in anatomy, surgery, and botany; also distinguished himself as a dentist and oculist, and discussed the whole range of surgical topics from the least to the greatest.

Bilguer (1720–1796), of Chur, became surgeon-general in Berlin, and performed the first resection of the wrist in

1762; he was an opponent of amputation, which at that time was altogether too frequently practiced.

Von Siebold (1736-1807) was the founder of an institution for surgical instruction, where, for the first time in Germany, surgery was taught clinically. He became one of the most famous teachers, and was first in his native land to perform the operation of symphysiotomy, so recently revived

The greatest German surgeon of the eighteenth century, however,—one eminent both as writer and operator, —was August Gottlieb Richter (1742-1812), of Zorbig, a descendant of a ministerial family, who wrote a famous work on hernia, and greatly improved all branches of surgery; he it was that enunciated the principle of dressing wounds " quickly, easily, and rarely."

Among English surgeons of the century must be mentioned, first of all, Cheselden (1688-1752), whose name is inseparably connected with anatomy and pathology as well as surgery At first a warm advocate of the high operation for stone, his dexterity in lithotomy excited the wonder of his contemporaries. He published a treatise on anatomy, and one on the suprapubic section.

Alexander Monro, Sr. (1697-1767), of Edinburgh, was also eminent in both anatomy and surgery, and contributed more than any other one man to the success and reputation of the Scottish medical school. His sons, Alexander and Donald, and his grandson, Alexander (3d), were equally celebrated in anatomy.

Charles White, of Manchester, is generally credited with having performed, in 1768, the first subperiosteal resection of the head of the humerus, although, as a matter of fact, this was not done until 1774, and then by Bent, of Newcastle. He also performed resection of the hip-joint upon the cadaver—another of the same name, Anthony White, having done the operation on the living subject in 1721. He invented the method of reducing dislocation of the humerus with the foot in the axilla,—a procedure that is ordinarily ascribed to Sir Astley Cooper; also operations for false joint by the removal of the involved surfaces of the bone.

It will be seen that the excision of the joints was peculiarly an English method, the elbow-joint having been first excised in 1758, by Wainman, and the knee-joint by

FIG. 32.—WILLIAM HUNTER, M.D., F.R.S.
(From a steel engraving by J. Thomson, made from a painting by Pyne.)

Filkin, of Northwich. The man who permanently attracted the attention of surgeons to these new operations was Henry Park, a bold surgeon, who wrote in 1782. The merits of these methods were then soon forgotten, however, and were revived in the present century by Liston and Syme.

One of the best-known London surgeons was Percival Pott (1749-1787), who became especially eminent through his studies upon hernia, spinal disease, and diseases of the bones and joints; his complete chirurgical works appeared in London in 1771.

William Hunter (1718–1783), of Scotch parentage, originally a theological student, and a pupil of Cullen, went to London in 1741, began to lecture on anatomy and surgery in 1746, and soon acquired a great reputation as a surgeon, obstetrician, and anatomist. He achieved enormous success in practice, and spent £100,000 upon his house, library, and private collections. The latter now form the Hunterian Museum in the University of Glasgow. His magnificent plates illustrating the gravid uterus required the labors of twenty years and appeared in 1774.

John Hunter (1728–1793), younger brother of William, enjoyed even greater reputation than the latter. He was a pupil not only of his brother, but also of Cheselden and Pott. Beginning the practice of surgery in 1763, he became surgeon to St. George's Hospital in 1768, and Surgeon-general of the English forces in 1790. So memorable were the labors and services of this man that at the Royal College of Surgeons, of London, there is given annually an "Hunterian Oration," intended in some way to commemorate his labors or to draw some lesson from his life and work. To do justice to John Hunter would require a volume, hence we must at present dismiss the subject with this brief reference.

Almost equally famous as a surgeon, though by no means such an omnivorous student as Hunter, was Benjamin Bell, of Edinburgh, who died in 1806. He employed tubes of lead and silver for the purpose of drainage. Sir Charles and John Bell, also of Edinburgh, are eminent names pertaining to the latter part of the eighteenth and first part of the nineteenth century. The latter was Professor of Anatomy, Surgery, and Obstetrics, a busy practitioner, a fertile writer, and not only one of the most successful operators of his day, but an excellent classical scholar; his *Principles of Surgery* appeared from 1801 to 1807. Sir Charles, who died in 1842, belongs more to the present century, but was equally distinguished as an operator, surgeon, and writer, and best known, perhaps, for his *Bridgewater Treatise on the Hand*.

Among the Dutch an eminent surgeon was Peter Camper (1722–1789), who, in order to acquire manual dexterity, learned to use various mechanical tools. He was a

FIG. 33.—JOHN HUNTER.
(From a steel engraving by G. H. Adcock of a painting by Sir Joshua Reynolds.)

fruitful author, and did not consider it beneath his dignity to write a treatise about the best form of shoes, published in Vienna in 1782, but recently translated and republished in England as something new. Sandifort, of Leyden, discussed ruptures, dislocations, etc., and reported the first observation of downward dislocation of the femur.

As already noted, the surgeons of the eighteenth century were often obstetricians,—William Hunter conspicuously. The most important obstetrician of his time was William Smellie (1680–1763), of London, who invented numerous instruments, wrote a large treatise on the theory and practice of midwifery, and greatly advanced our knowledge of deformed pelves. He was the first to distinguish one diameter from the other, and to point out the importance of cephalic version and version of the breech. Parenthetically, it may be remarked that William Hunter, great as he was, was the uncompromising foe of instrumental midwifery, and was in the habit of showing his forceps, covered with rust, as evidence that he never resorted to such aids. A rival of Smellie and Hunter was Thomas Denman (1753–1815), best known, perhaps, because of his demonstration of the portability of puerperal infection.

The researches of anatomists during the eighteenth century were, for the most part, directed toward the minute, more difficult, and less striking parts, and to increased thoroughness and accuracy of description. Microscopical anatomy suffered a relative quiescence. Pathological and general anatomy, which were destined to control the medicine of the succeeding century, were newly created and not yet regarded as sciences by themselves, but merely as special branches. The most important feature was the revival and more accurate study of experimental physiology, which had been scarcely resorted to since the time of Galen, except for Harvey's discoveries. This revival, which really seemed an epoch in the history of medicine, was effected by the great Haller (1708–1777), of Berne, —a man who really deserved the title of "Great," as he was a universal and indefatigable *savant*, possessed of thorough conscientiousness, marvelous capacity for work, great ingenuity, natural endowments, and an inextinguishable love for art and science; he was certainly one of the most versatile scholars and thinkers of any time, distinguished not only in his chosen field of medicine, but as a poet, botanist, and statesman. Like all Swiss poets, he never passed beyond the didactic and the homely in his versification. From his tenth year he wrote poems in Latin and German, and even when eight years old had made most extensive compilations from Bayle's dictionary. At fifteen he went to the University of Tübingen, where, in the second year of his sojourn, he disputed with one of his teachers. In 1725 he went to Leyden, where Boerhaave and Albinus found in him a most indefatigable follower. At nineteen he received the degree of doctor. In the excess of his zeal for anatomy he purchased for a considerable sum, from Albinus, half of a corpse, the other half of which his teacher had dissected; and, while in Paris, he even engaged in grave-robbing, and, being betrayed by his own carelessness, was compelled to save himself by flight. In many other States, and in more than one country, he studied with the best of teachers, lecturing at times himself. At the age of twenty-six he became professor and hospital director at Berne, and in 1752 published his famous researches on irritability. Three years later he accepted a call to Göttingen as Professor of Anatomy, Surgery, Chemistry, and Botany. He was the founder of a botanical garden; for many years was so busy that he slept and lived in his library; and, in spite of his enormous and unique correspondence with the *savants* of the world, he never left a letter unanswered. Strange to say, his permanent influence upon the practice of medicine was only indirect; and, although he was professor of surgery, and performed many vivisections, he was never able to persuade himself to perform a single surgical operation upon the living human being. He it was that introduced into Germany the use of the watch in counting the pulse. Like Hunter, Haller demands a special historian, and it is possible here to outline only a few of the services he rendered to medicine. He enriched the anatomy of the heart, of the brain and dura, and pointed out the venous nature of the sinuses; taught that the uterus should be regarded as a muscle; advanced the knowledge of the lymphatic system, and believed in and taught a developmental theory that every individual is descended or derived from a preceding one. In the mechanism of the heart his doctrine of irritability especially maintained itself. He administered the death-blow to the doctrine of vital spirits, and was, in fact, the father of modern nerve-physiology. His doctrine of irritability moved the minds of his century in a way that has no parallel, unless we compare it with the doctrine of Darwin. Glisson had established the general principles of irritability, and Haller followed, teaching it by the inductive method, and proving its existence by experiments, —proving, moreover, that it is a peculiarity of the muscular substance and not governed by ordinary sensation. His researches deserve the more credit because he lacked modern aids to physiological study. The first physiological institute was founded in Breslau by Purkinje, some fifty years ago. Haller had no such opportunity; even his successor, the great Müller, possessed no such advantages. The profound impression made by Haller's teachings may be measured by the number of his supporters and opponents; he was a great man, second only in wide-spread influence to Boerhaave, and one who left a more lasting impress upon the world than even the latter.

The two best known of Haller's opponents were: Wolf (1733–1794), of St. Petersburg, who regarded each generation as an actual new creation, and was the first to teach the doctrine of the blastodermic membranes; and Blumenbach (1752–1840), of Gotha, who did great service by investigations in general anthropology, of which he was, in fact, the founder, and whose researches in comparative anatomy and the history of development have rendered him famous.

Of the famous anatomists of the century may be mentioned Sömmerring (1755–1830), of Frankfort,—the first to distinguish the facial and auditory nerves from each other, and whose published works are well known, because of the beautiful illustrations furnished him by the well-known artist, Koeck.

The ablest French anatomist of the century was Winslöw (1669-1760),—a man of Danish birth, but who

Fig. 34.—J. F. Blumenbach.
From an old steel engraving.

became a professor in Paris, and is best known by the foramen named for him. There were, also, Portal (1742-1832). physician to Louis XVIII. who wrote a famous history of anatomy and surgery; and Vicq d'Azyr (1748-1794). known equally well for his labors in the department of anatomy, especially of the brain, nervous system, and the vocal organs. Bichat (already mentioned) would deserve to be placed at the head of French anatomists were it not for his superior rank in clinical medicine.

The founder of pathological anatomy as a science was Morgagni, born in 1682, in Forli, Italy,—a pupil of Valsalva, and, at the age of nineteen, the assistant of the latter. It was not until his seventy-ninth year, after he had published several works, that he allowed his famous work on pathological anatomy to appear. This is the historical classic, *De Sedibus et Causis Morborum*, published in Venice in 1761. Its famous author did not cease work, even when he became blind, and to him we owe the maxim that observations should be "weighed, not counted." He was very versatile, and well informed in all branches of science and literature, and possessed a remarkable memory; likewise was the first to devote attention extensively and thoroughly to the anatomical products of common diseases, since, before his time, little had been regarded but rare discoveries in the body. He also called attention to the important bearing which the history of the disease has toward its products, and found his discoveries of advantage, even when they were unable to promote the cure of disease, because of the light which they threw upon physiology and normal anatomy, and because they prevented incurable patients from being continually tormented with drugs intended to cure them; also because pathological investigations alone could settle disputes in diagnosis and matters of honor among physicians. He died in 1772.

Morgagni's legitimate successors in Great Britain were Baillie (1761–1823), a son of John Hunter's sister, and Sir Everard Home,—Hunter's brother-in-law,—who became professor in the Royal College of Surgeons, and was intrusted by Hunter with the work of describing his collection. Home, however, in a most discreditable way, burned several volumes of Hunter's own descriptions, in order to appropriate to himself the sole credit of the work. He has gone down to fame especially because of his book on the prostate.

One of the most notable events in the history of medicine was the introduction of the systematic practice of *preventive inoculation against small-pox*. It is so generally taught that this is entirely due to the efforts of Jenner —or, rather, we are so often allowed to think it, without being taught otherwise—that the measure deserves an historical sketch. The communication of the natural disease to the healthy, in order to afford protection,—or, in other words, the communication of small-pox to prevent the same,—reaches back into antiquity. It is mentioned in the Sanscrit Vedas as performed by Brahmins, who employed pus procured from small-pox vesicles a year before. They rubbed the place selected for operation until the skin was red, then scratched with a sharp instrument, and laid upon it cotton soaked in the variolous pus, moistened with water from the sacred Ganges. Along with this measure they insisted upon careful hygienic regulations, to which, in large measure, their good results were due. Among the Chinese was practiced what was known as "pock-sowing," and ten centuries before Christ the Celestials introduced into the nasal cavities of young children pledgets of cotton saturated with variolous pus. The Arabians inoculated with needles, and so did the Circassians, while in North Africa incisions were made between the fingers, and among some of the negroes inoculation was performed in or upon the nose. In Constantinople, under the Greeks, the custom had long been naturalized, and was practiced by old women, instructed in the art, who regarded it as a revelation of Saint Mary. The first accounts of this practice were given to the Royal Society by Timoni, a physician of Constantinople, in 1714. The actual introduction of the practice into the West, however, was due to Lady Mary Wortley Montagu, who died in 1762, and who was wife of the English Ambassador to the Porte in 1717. She had her son inoculated in Constantinople, by Maitland, and on her return to London, in 1721, her daughter also was inoculated. During the same years experiments were undertaken by Maitland upon criminals, and, as these turned out favorably, the Prince of Wales and his sisters were inoculated by Mead. The practice was then more or less speedily adopted on this side of the Atlantic, but suffered occasional severe blows, because of unfortunate cases here and there, such as never can be avoided. The clergy, especially, using the Scripture, as designing men can always do, became warm opponents of the practice, and stigmatized it as an atrocious invasion of the divine prerogative. Nevertheless, in 1746 the Bishop of Worcester recommended it from the pulpit, established houses for inoculation, and thus made it again popular. In Germany it was generally favored, and a little later came into vogue in France and Italy. In 1757 Robert Sutton, near London, professed to have made fifteen thousand inoculations without a single fatal case; he kept his patients on a strict diet for nine days, then inoculated with the smallest possible quantity of virus. The operation was not prohibited in England until the year 1840, although it involved much greater dangers than vaccination with cow-pox.

The first inoculation with cow-pox seems to have been performed in 1774 by a farmer of Gloucester, named Jesty, though the pioneer in the extensive and general introduction of this method was Edward Jenner (1749–1823), of Berkeley, in Gloucestershire, who, therefore, is generally known as the "Father of Vaccination." The son of a clergyman, he began early the study of medicine and surgery, and during his apprenticeship received from a milkmaid information of the protective power of cow-pox against variola, as established by popular observation. (Sutton and others had proved that inoculation of *sheep*-pox was not efficient.) This communication so struck Jenner as a means of affording protection to the whole human race that the subject never afterward left his mind. In 1770 he became a pupil of John Hunter, and when he communicated to him this idea the great surgeon said: "Do not think; investigate!" Accordingly he went to Berkeley and performed the little operation which has made him famous; and from 1778 until 1788 he communicated to Sir Everard Home such observations as he had made. But the first vaccination was performed in 1796,

FIG. 35.—EDWARD JENNER, M.D.
(From a steel engraving by E. Scriven made from a painting by J. R. Smith.)

upon a boy, with matter from the hand of a maid who had contracted cow-pox in milking. In 1798 he published his memorable work, and afterward removed to London. He died full of fame and honor, in his native place, having received rewards from the government amounting to one hundred and fifty thousand dollars, besides being made an honorary citizen of the city of London. The subsequent wide-spread practice of the method, and the formation of societies for the promotion of vaccination are matters of recent history.

The first vaccinations in the United States were performed by Doctor Waterhouse, Professor of Medicine in Harvard College, in 1800, upon four of his own children. The transmission of humanized virus through the system of the cow, and its subsequent employment in vaccination of human beings, was first practiced by Troja (1747–1827), of Naples, shortly after the introduction of human vaccination; but in 1810 this was prohibited in Italy. Compulsory vaccination was first extensively introduced in Germany in 1807; in England it was first legalized in 1827. The occasional temporary character of the protection thus afforded was first taught by Elsässer in 1814. Schoenlein was the first to call attention to the distinction between variola and varioloid.

Another matter in which the eighteenth century witnessed great reform was the *treatment of the insane*, which continued in very bad condition until toward the close of the century, when a movement for improvement began. From and after this lunatics were liberated from their fetters and from the hands of brutal keepers, and regarded as actually ill, while so-called schools of psychiatry were founded. While the first impulse in this direction was given by Lorry, the true reformer was Pinel, already mentioned, who did away with corporeal punishment and abuse, separated the insane from convicts, limited the employment of drugs and especially venesection, placed the unfortunates in special institutions under the charge of physicians, and classified patients according to their symptoms. Yet, in spite of his humane teachings, lunatics were found incarcerated in cages in some of the French cities as late as 1834. Pinel was followed by Esquirol (1772–1840), who in 1818 established the first clinic for mental diseases.

It is well known what a conspicuous part *public baths* played in the social life of the ancient Greeks and Romans, but the first public resort for sea-bathing was established in Germany in 1794. The cold-water epoch of this century, however, began with the researches of Hahn (1696–1773), a Silesian, who introduced a systematic and almost exclusive hydrotherapeutic method. The modern method of using cold water as an antipyretic agent was first employed in England, in 1797, by Currie, who originally was an American merchant. In France the method found little sympathy, but it made its way even to Spain later, where it was adopted by the famous Sangrado, who is well known to readers of *Gil Blas*.

CHAPTER IX.

THE AGE OF RENOVATION (*continued*).—The Eighteenth Century; General Considerations. Foundation of Learned Societies, etc. The Royal College of Surgeons, 1800; the Josephinum, 1785.—The Nineteenth Century. Realistic Reaction Against Previous Idealism. Influence of Comte, of Claude Bernard, and of Charles Darwin, 1809-1882. Influence Exerted by Other Sciences.—*Theory of Excitement:* Roeschlaub.—*Stimolo and Contrastimolo:* Rasori, 1762-1837.—*Homœopathy:* Hahnemann, 1753-1843.—*Isopathy, Electrohomœopathy* of Mattei.—*Cranioscopy, or Phrenology:* Gall and Spurzheim.—*The Physiological Theory:* Broussais, 1772-1838.—*Paris Pathological School:* Cruveilhier, 1791-1873. Andral, 1797-1876. Louis, 1787-1872. Magendie, 1783-1855. Trousseau, 1801-1866. Claude Bernard, 1813-1878.—*British Medicine:* Bell and Hall, Travers, 1783-1858.—*Germany, School of Natural Philosophy:* Johannes Müller, 1801-1858.—*School of Natural History:* Schönlein, 1763-1874.—*New Vienna School:* Rokitansky, 1804-1878. Skoda, 1805-1881.

THAT the eighteenth century, up to its close, was the golden age of medicine, is due to the prevalence during that period of a strong idealistic undertone, as a result of which any learned occupation caused the scholar to be held in higher esteem than is the case even to-day. Medicine was then regarded as *a conscientious vocation* and not as a mere business or trade; indeed, general scientific knowledge more widely prevailed among the better class of the profession, and there was much less of that one-sided, narrow education that obtains to-day. The profession, moreover, was not overcrowded; physicians were neither too few nor too numerous, consequently their social position was higher. Again, the relations between doctor and patient were more intimate, most practitioners being of the type described as "family physicians," and those possessed of the doctorate degree ranked among the gentry rather than as artisans. They were, for the most part, fully devoted to their calling; moreover, the State took greater care to protect the people, so that it became dangerous for strolling vagabonds and imposters to attempt to trifle with human life and excite the vulgar to the prejudice of scientific knowledge.

(230)

The pursuit of anatomical studies was now facilitated, despite the fact that students were frequently compelled to take long journeys in order to obtain the "material" therefor. In the early part of the century so great was the lack of dissecting material that the great Haller while in Paris was compelled to purloin his cadavers, and ultimately, on discovery of this fact, to fly for his life; Hoffmann was only able to make twenty dissections during twenty-four years; even in the middle of the century there was only one dissection annually in Halle; up to 1712 there had been only three dissections in a score of years—though now subjects can be had there in abundance at a ridiculously low figure; cadavers were extremely scarce in Vienna as late as 1765; and for a long time the only places in London where the study of anatomy could be legally pursued were the College of Physicians and the College of Surgeons, and the trouble that hampered John Hunter in this direction is historical. The crime of "Burking" became known in Hunter's day. Murder was committed, and the victim sold for purposes of dissection—for at this time, as "body-snatching" was a necessity, those that purchased cadavers asked few questions, and the fees paid were, of course, high.

The first clinical institution in Austria was organized in Vienna, in 1754, by Van Swieten, though there was an "ambulatory clinic" (out-patient department) in Prague nine years before. During the century, however, hospitals were everywhere in bad condition. In the Hôtel-Dieu, at Paris, several patients—even as many as six—were sometimes put in the same bed; the convalescent and the dying found themselves thus associated; in Vienna the Allgemeines Krankenhaus was composed of seventeen hospitals that subsequently were amalgamated into one. In London numerous hospitals were founded, and as the medical staff of each became eminent they attracted numerous pupils; but later it became necessary to relieve the hospital wards, and private institutions for instruction were established by popular teachers, the most celebrated being the "Windmill Street School of Anatomy," founded by William Hunter about 1770, and the private school of Sir William Blizzard, which, established in 1780, developed, five years later, into the London Hospital Medical School.

While few, if any, of the lectures were compulsory, particularly in the natural sciences, even more attention than now was bestowed upon the accessory branches; botany, chemistry, and natural history were the recreation of many students and physicians. Pupils enjoyed the privilege of studying what they pleased—as they do practically to-day in the Portuguese University of Coimbra,—and professors exercised to the utmost their individuality in teaching. In Spain natural sciences found no admission, and even so late as 1770 no instruction in these branches was given, as they were regarded as dangerous to the purity of the faith; mineralogy for mining purposes was an exception, for even the most faithful Catholic needs money.

At the universities medical students were not permitted to go out without their scholastic cloaks,—a regulation that still obtains in Spain. That the number of students has enormously multiplied may be seen from the fact that the little University of Giessen, with scarcely any medical school at all, has always more students than had Halle in the days of the famous Hoffmann. In the middle of the last century Würzburg had at one time but *three* medical students, while to-day it has in the neighborhood of five hundred. Even then it was complained that, on account of the number of students, there was an educated proletariat arising, and in 1791 it was proposed, in Austria, that the rush for study should be repressed.

Among the Continental students the revels and bad behavior of past centuries were not to any great extent corrected; fights and debauchery were very common, and all sorts of orgies and bacchanals prevailed. The professors were, in large measure, independent of the State, and a single individual often represented a number of branches now taught by special chairs. When indisposed to lecture, they simply posted upon the blackboard: "*Hodie non legitur*," and this was the end of the matter. In 1777 Vienna had one hundred and forty-seven medical teachers, and in Germany there were two to every thirty-nine students. That in the last century one man often accomplished more than a great number of average teachers do to-day is amply demonstrated by the lives of Boerhaave, Haller, and others. Then, too, the Latin tongue was generally employed for purposes of instruction, though surgeons, for the most part, lectured in the vernacular; Cullen, in 1770, was the first in Great Britain to deliver purely medical lectures in English; and as the clergy gradually retired from the ranks of the profession, Latin more and more fell into disuse. Strange to say, as the clerical influence waned, the Jews began to enter medicine, the movement beginning about 1791, in France, under the promulgation of "civil equality" ideas; previously the Hebrews had been an almost universally suppressed people, and in Berlin were permitted to enter and leave the city by only one gate, and were forbidden to learn or write pure German, in consequence whereof their dialect was an Hebraic-Teutonic jargon, that even to-day prevails in some portions of western Europe. Educated Jews were few in number, since attendance upon universities was ordinarily denied them, although long before they had been admitted at Salamanca, Toledo, Salernum, and Montpellier. In Austria the prohibition was not removed until 1789, and even then, so bitter was the prejudice against the Semitic race, the clergy vigorously protested. It was the same clerical body that, in 1667, protested with the greatest vehemence against allowing Hebrew physicians to pass through the gates of Würtemburg without paying toll, declaring that it was "better to die with Christ than be cured by Jews, who were aided by the devil."

Professors were often attached to the courts of their various sovereigns, and at one time the French court possessed a faculty of forty-eight physicians, surgeons, and apothecaries, the first two physicians being required to attend every morning when the king arose; hence originated the titles, still known in Germany, of "Hofrath" and "Geheimrath."

Medical fees, as a rule, were very small, though there were exceptional instances in which enormous sums were bestowed: Joseph II, of Austria, gave Guerin, who was summoned from Paris in consultation, an honorarium of 171,000 marks and made him a baronet. Taking all things into consideration, the income of the average practitioner in the eighteenth century would be in the neighborhood of $1000, which, however, was equivalent to three times that amount to-day. Fothergill, whose highest income in a single year was $25,000, bequeathed to the poor of London $1,000,000; Sir Astley Cooper had a yearly income of from $75,000 to $100,000, but it may be remembered that his practice during the first year netted him just $26, and that it was four years later before his income reached the sum of $500.

The physician of the last century was, at least, on occasions of moment, very different from other men, and to be recognized by his dress. A cap was placed upon his head when he graduated, in recognition of the fact that physicians at an earlier period belonged to the learned or clerical profession; and in later life he wore a purplish or scarlet cloak (to distinguish him from lawyers, whose professional color was yellow, and from theologians, who then, as now, sported the sombre black). The regulation full-dress costume of the English physician of the last century demanded a well-powdered wig, silk coat, knee breeches with stockings, buckled shoes, lace ruffles, cap, and gold-headed cane, to which, in cold weather, was added a muff —to preserve his delicacy of touch.

Surgeons were still strictly separated from physicians, even in education; nor were they esteemed as equal in rank, until the French Revolution brought about the doctrine of civil equality; perhaps this is one reason why this branch of the medical art made less conspicuous progress until recent times. The change was brought about, in France, by the abolition of eighteen universities and fifteen colleges of medicine, the Royal Society of Medicine (founded in 1776), and the Academy of Surgery (founded in 1731); but by this abolition charlatanism acquired such speedy control that the arrangement was soon abandoned. Thus it came about that surgical instruction was given in special institutions or in the universities, and the conditions of instruction finally improved. When the College of St. Côme was abolished in 1753 the Société de Chirurgie, founded in 1731, became the Académie de Chirurgie; and, when the French Academy was formed in 1795, the Académie was merged into its medical department. The École Pratique, where Desault and Chopart taught, was established in 1750 for the practical education of surgeons. In England the Royal College of Surgeons was not incorporated until 1800. In Austria, in 1785, the Josephinum was opened by Joseph II, who also erected permanent military hospitals in Prague, Brünn, Milan, Mantua, Pesth, Olmütz, etc.; he also created the "Joseph's Akademie" in order to educate military surgeons and thus overcome the defects of army surgery; the Josephinum unquestionably exerted great influence in elevating the social and military position of army surgeons and attained historical importance after Brambilla compelled the recognition of surgeons as social equals of other members of the medical profession. As the result of these improvements, the various armies of Europe were soon furnished with better medical officers. Prior to this, too, the field hospitals had been as badly mismanaged as their civil prototypes, and the substitution, in 1793, of movable hospitals, as suggested at the close of the sixteenth century by Henry IV, of France, was scarcely an improvement. The whole system suffered from perpetuation of the dual and distinct functions of the physician and the surgeon, to destroy which was a part of the design of the Josephinum. How unpleasant was the position of the army surgeon up to this date may be inferred from the fact that in 1758 one was subjected to corporeal punishment at the command of his colonel, and that a general upon his death-bed could leave orders that fifty blows be given each of his medical staff in case the post-mortem disproved the diagnosis.

In Austria, at the beginning of the Seven Years' War, all military surgeons of the Protestant faith were compelled to become Catholics or leave the service. The condition of the wounded soldiers was as deplorable as can well be imagined; but upon this subject I cannot dwell.

The tendency of the nineteenth century seems to be a continuation, and, perhaps, in some respects, an exaggeration, of the condition obtaining in France during the previous century; in other words, the world has become practically an enormous school of pathological anatomy and diagnosis,—a school inaugurated by Bichat, as representing so-called scientific or exact medicine. Philosophically this has been a century of reaction against the idealism of the preceding age; it places the individual, rather than the idea, in the foregound. The mutual influence of medicine, philosophy, and the natural sciences is less conspicuous now than formerly. Recent philosophers who have exercised the greatest influence are: Schelling, who held to the equality of the real and the ideal; Hegel, whose supreme principle was absolute reason, of which religion was regarded as a representation; Hartmann, whose philosophy of the "unconscious" depends largely upon the results of natural sciences, embraces Darwinism, and is, in many respects, an extension and completion of

Schopenhauer's pessimism and doctrine of the soul. But one who has exercised still more influence upon our profession is Comte, whose positivism contrasted strongly with the idealism and atheism of Schelling, and who required only this of philosophy,—namely, that it should work out the general ideas and results of other sciences; his most important follower was Claude Bernard, and upon these two the whole exact school of France is based. But the most influential philosophic doctrines of this or any other century have been those emanating from Charles Darwin, Herbert Spencer, Ernst Haeckel, Alfred Wallace, and their contemporaries and followers. Darwin (1809–1882) was the grandson of Erasmus Darwin, already mentioned, and his *Variation of Animals and Plants Under Domestication*, *Origin of Species*, and *Descent of Man* have found a place in all modern languages. The system known by his name is the pure science of nature, is founded upon scientific investigation, and by its merits alone has found almost universal acceptance; it has been added to and further elucidated by the efforts of Haeckel and Spencer.

When it is declared that medicine of the present is influenced by no system, it is speedily found, on critical analysis, that this is an error. It necessarily follows the realistic and materialistic as readily as it did the teachings and doctrines of natural philosophy; and, in consequence, "medical thought," so called, is just as one-sided to-day as at any time in the history of the art. The watchword of to-day, "natural specific tendency," veils, but does not take away, its philosophic principles, and so our ridicule of earlier medical systems is quite unjustifiable. A modern historian aptly remarks that the medicine of the present "embraces nothing but a theorem of investigation by the senses."

Discoveries in botany, the result of better knowledge of natural history and more accurate habits of study, have influenced modern progress not a little; have led to better classification and broader knowledge. The natural system of de Candolle (1778–1841), of Geneva, and of Endlicher, of Vienna, called into existence the so-called natural historical school of medicine; the researches into plant-cells by Schleiden and Baumgärtner, and the almost contemporaneous discovery of animal cells by Schwann became, in course of time, the origin of recent cellular pathology; then came microscopic botany, and the influence of the lower fungi in the production of fermentation and putrefaction.

Similarly, too, the laws of physics have been shown to have an inseparable connection with anatomy and physiology, and their study has become a most important aid in the experimental researches of to-day; through Helmholz they brought in the ophthalmoscope; thermal electricity, for the discovery of which medicine is indebted to Seebeck; a better knowledge of optics, thanks to Fraunhofer, who was equally expert in electricity; spectrum analysis, invented by Kirchhoff; and the varied efforts of Faraday, Graham Bell, Thomas Alva Edison, and Daguerre, the latter better known for his invention of photography. Finally, medicine is immeasurably indebted to Tyndall and Huxley for their teaching of the correlation and conservation of energy.

Chemistry also has performed its share, and, as applied to physiology, is a discovery almost wholly within the present century. The new nomenclature serves a practical purpose in that it is now possible to portray chemical combinations and isomerism in a graphic, and, at least, semi-comprehensive way. Among the chemists may be specially mentioned Bertholet, whose laws are as well known as they are succinct; Humboldt; Berzelius; Dumas; Chevreuil, who recently died at the age of almost one hundred years; Magendie; Orfila, the toxicologist; Gmelin, eminent in physiological chemistry; Rose, perfecter of organic analysis; Wöhler, who first made organic alkaloids; Bunsen; Sir Humphry Davy; Marsh; Faraday; Graham; Young, who first showed the industrial value of coal; and Gay-Lussac.

Upon medicine, zoölogy also, with comparative anatomy and physiology, has had a wonderful influence; here may be noted the names of Cuvier, Oken, Bilharz, Brehm, Wagner, Leuckart, Richard Owen, William Carpenter, and last, but by no means least, Thomas Huxley.

But perhaps the most significant feature of the age has been the wonderful development of scientific associations and the publication of medical and scientific literature. Whether these have yet reached their climax is perhaps an open question, but the consequent widening circle of readers, as well as of writers, seems to imply that there will be for a long time to come no lack of activity in this direction. In the United States more than in any other country medical societies and associations innumerable have sprung up, and to such a degree that (in the eastern States at least) there are few counties that cannot boast of a medical organization.

During the present century foreign universities have decreased in number, partly owing to consolidations and partly by surrender of charters; for instance, the old University of Ingolstadt was united with that of Landshut, and in 1827 was removed to Munich; in 1816 the University of Würtemburg was united with that of Halle; the University of Bonn was abolished in 1792, but revived in 1818. A few new universities, like that of Zürich, have been founded. In the quaint old town of Prague the old German university was, in 1883, divided, and there now exist in that city two universities side by side, in one of which German is spoken, in the other Bohemian.

It will thus be seen that the nineteenth century is essentially an era of modern science, with whose dawn was sounded the death-knell of the "demon of disease" and his twin brother "visitation." In 1801 the first experiment in steam-navigation took place upon the Thames. In 1807 the slave-trade in England was abolished by Parliament. The theological part has entirely faded out of medicine; and the era of accurate scientific experimentation which long since dawned, is now, so far as we can see, at its height, since it is difficult to conceive of much improvement upon its methods under existing conditions, or of greater enthusiasm than has been already manifested.

Now, regarding some of the systems and theories of this age. The systems of the past have been more or less long-lived,—as, for instance, those of Dogmatism and of Galen,—while as we come closer to the present they become more ephemeral. Those of the early part of the present century took root in the soil of the eighteenth,—for instance, the so-called *theory of excitement* of Roeschlaub (1768–1835), which endeavored to mold into one the Brunonian errors and the fancies of Schelling. According to it, life depends upon irritability, but is inherent in the organism as an independent feature; so it recognizes both irritability and solidism, while Brown considered the former alone, adding, as an after-thought, a chemical or qualitative potency (oxygen), in order to account for alterations of quality. Roeschlaub inclined first toward natural philosophy, then, owing to an inherent theological and polemical bias (he was originally intended for the church), to mysticism and theosophy; finally, with a courage almost unexampled, he upset all his former teachings by admitting he was mistaken. To him was opposed Hufeland, who wrote on the *Lengthening of Life*, was noted for a warm and benevolent heart, and possessed no small penetration, as is evidenced by his aphorism, "Successful treatment requires one-third science and two-thirds '*savoir faire*.'"

Stimolo and *contrastimolo* were titles applied to a theory advanced by Rasori (1762–1837), of Milan, that combined Methodism with Brunonism; by Baas it is characterized as a "genuine blot upon the human heart beyond any other of the various systems." Long centuries of experience and the conclusions of great and venerable minds may go for naught, as Rasori abundantly demonstrated. The theories of Brown were then taught as his own to his classes in Pavia, showing he was not above plagiarism; his *stimolo* corresponded to the sthenic diathesis devised by Brown, while his system consisted of an endeavor to make a diagnosis by watching the effects of drugs. Bleeding was held to be the best measure; if it did the patient good, the sthenic diathesis was assumed; if it made him worse, the asthenic was certain. He gave enormous doses of powerful drugs—sixty grains of gamboge, and from two to three ounces of saltpeter in a single day. Is it strange that homœopathy or any other heterodox system sprang up in the midst of such measures? It is an old saying that there is no folly which will not secure a following; and, strange to say, Rasori had a numerous and an eminent one.

As just intimated, Homœopathy was the natural reaction against such heroic measures; in the rebound the other extreme was reached, even to practical therapeutic nihilism. Now, instead of venesection and drastic medication, came the theories expounded by Hahnemann (1753–1843), which denied disease, admitting only symptoms. This apostle of homœopathy was the son of a porcelain-painter in Meissen; he studied in Leipzig and in Vienna, and later practiced in various cities, including Dresden and Leipzig. "*Similia similibus curantur*" was not original with him, as it long before had been formulated by Hippocrates, and later by Paracelsus. Of the life and labors of Hahnemann, much might be told; but this is not the time or place to go into the subject.

An offshoot of homœopathy, which demands only the harshest criticism, is Isopathy,—perhaps the filthiest theory ever invented,—according to which like is to be cured by like, and to such an extent that small-pox is to be treated by variolous pus, tape-worm by the ingestion of the proglottides, etc.

Another of the rankest of fraudulent outgrowths is the so-called Electrohomœopathic system of Count Mattei, who prates of "red," "blue," and "green" electricity,—a theory that, in spite of its utter idiocy, has attracted a con-

FIG. 36.—SAMUEL HAHNEMANN.
(From a steel engraving.)

siderable following and earned a fortune for its chief promoter.

Another of the vagaries of the earlier portion of the present century, and that still survives, in a weak way, is Cranioscopy, or Phrenology. Gall expounded his doctrines at Vienna as early as 1796, but, being expelled, went to Germany, where he was joined by Spurzheim, who, though much more of a student and scientist, accepted the doctrine of the former with enthusiasm; and it was chiefly due to the efforts of Spurzheim that phrenology was introduced into England, and later (1832) into America. Gall assumed to locate twenty-seven different organs alongside of each other in the brain, and held that external markings on the skull were guides to the development of the various parts. Every neophyte in anatomy knows how little foundation there is for such a doctrine, but for a time it attracted great attention, and there are to-day certain men and women who make their living out of this imposition.

The Physiological Theory of Medicine was originated by Broussais (1772–1838), and combined the views of Pinel and Bichat with the "sympathetic" view of Hoffmann, the "concealed inflammation" of Stoll, and the theory of inflammation held by Marcus. Broussais had been a pupil of Bichat. In 1814 he began hospital teaching, and in 1831 was made professor. Personally very vain, quick-tempered, even belligerent, as a therapeutist he was a man of routine. He was, perhaps, best known shortly before his death, when delivering lectures on phrenology. According to him, life depends upon external irritation, produced by heat, which excites new chemical processes, while these in turn stimulate regeneration, assimilation, as well as contractility, and sensibility. When the functions supported by heat cease, death ensues. Health depends upon moderate action of external irritants; disease, upon either their weakness or their extraordinary strength. He saw nothing ontological about disease. In therapeutics he admitted the healing power of Nature, but regarded the physician not as a minister, but as a lord of Nature. Febrile and inflammatory diseases were all treated by the withdrawal of nourishment, carried to the extreme. His most powerful antiphlogistic treatment consisted in the application of leeches to the abdomen, and to robust individuals he applied from thirty to fifty at once It is not, then, to be wondered at that, in consequence of his so-called "hirudinomania," leeches became very scarce In the year 1833 forty-one million five hundred thousand leeches were imported into France, while in 1824 one-twentieth of this number sufficed to supply the demand. Even in cases of worms, the abdominal integument had to pay its blood-tribute, particularly if enteritis prevailed. He only allowed a spare diet of mucilaginous and acid drinks. In mercurial France and Italy he gained numerous followers, but they were few and far between in practical, hard-headed Germany and England. His best follower was Bouillaud (1797–1881), who adopted the symptomatic nature of fever and the sanguinary therapeutics of his master, but used the lancet more than the leech. As the homœopaths regard Hahnemann, so Bouillaud looked up to Broussais as the Messiah of medicine and science, which, as Baas says, were "already greatly overstocked with Messiahs."

Contemporaneous with the school of Broussais, and its antagonist in all respects, was the Paris School of Pathological Anatomy and Diagnosis, which has given tone to all medical art. It made it the duty of the physician to search for changes in the human body, to investigate the local products of disease, and assigned to medicine the duty of removing these products. The tendency of its teaching was to treat the patient rather as a living cadaver than as a sentient being endowed with vital forces, and the charge which Asclepiades once falsely made against Hippocrates was revived upon new grounds. Kratzmann wrote some years ago: "In France every one experiments on the sick, less to attain the best method of cure than to enrich science with an interesting discovery and to advance the accuracy of diagnosis by some new physical sign." The seductiveness of this system promoted still more one-sidedness, which finally almost attained the belief that the science of medicine really originated in the Anatomical School of Paris.

The forerunners of this school were Bichat and Pinel, and its proper founders were Corvisart, Dupuytren, and Laënnec. There was also Bayle, who was first to apply the ear to the thorax in disease of the heart, and thus became the predecessor of Laënnec and Chomel. He was the godfather of typhoid fever, and from being a famous clinician became later a great pathologist. The most celebrated adherent of the method, however, was Cruveilhier (1791–1873), professor first in Montpellier and then in Paris, who revived the Anatomical Society founded by Bichat, and wrote his first essays as the result of Dupuytren's advice; finally, there came from his pen the famous treatise on *Pathological Anatomy*, with its magnificent plates,—a work begun in 1830 and not fully completed until 1864. Like Morgagni, he associated general and pathological anatomy with bedside observations; also established a class of inflammations to which belong gangrene and atony, and a certain class of neuroses and fevers, and endeavored to investigate the different steps in the development of lesions, not simply their final products. His teachings concerning pyæmia and phlebitis, which had been first studied by John Hunter, excited great attention, and he even came to the one-sided conclusion that "phlebitis rules the whole of pathology." He was the first to observe that its suppurative form does not occur primarily, but is secondary to coagulation of the blood.

The ablest representative of this school, and one who, perhaps, more than any other man, made Paris a Mecca to which foreigners made their pilgrimages, was Andral (1797–1876),—the son of a physician and the most noted and indefatigable investigator and thinker of his time. Between 1823 and 1840 were published the five volumes of his *Medical Clinic*, which made him famous. He taught, in opposition to Broussais, the existence of primary diseases of the blood, the so-called dyscrasiæ; made physiology subservient to pathology; was the creator of the chemistry of the blood; and in therapeutics was wedded to emetics and cathartics, ascribing little importance to abstraction of blood.

The first man to apply the Numerical Method to pathology, and who brought about the downfall of Broussais, was Louis (1787–1872), who had studied in Russia, but came to Paris while still a young man. He expressed his principle in the following words: "As often as I have formed an *a priori* idea and had afterward opportunity to prove the facts, I have invariably found that my idea was false. In pathology as well as in therapeutics numerical analysis is a useful practice. By numbers only can be obtained the frequency of conditions or this or that symptom; by a definite enumeration alone is it possible to utilize the special relations of age, sex, constitution of our patients, to settle that this or that symptom occurs so often in one hundred or one thousand cases." This system he applied to etiology, symptomatology, prognosis, therapeutics, and pathological anatomy. He discarded blisters and condemned large bleedings, but fell into other errors, carrying his numerical method to an unjustifiable extreme.

Next to Andral and Louis should be mentioned Magendie (1783–1855), Professor of General Pathology in the College de France, and physician to the Hôtel-Dieu, who was a representative of the new French medicine, and introduced experiments into both pathology and physiology; he was the pioneer in experimental pharmacodynamics, which occupies itself largely with alkalies, a large number of which he introduced into practice. He was a solid humoralist in pathology, a most accurate diagnostician, but (it is charged) "was too simple in therapeutics"! As a result of his intravenous injections of putrefactive material, he had the terms "pyæmia," "ichorrhæmia," and "metastasis" introduced into pathology.

Trousseau (1801–1866), of Tours, also became professor in the Paris Faculty, and rendered especial service in his studies of croup and the employment therefor of tracheotomy, though his chief fame rests upon his merit as a clinical teacher and the publication of clinical lectures which are still models in every way of accurate, forcible teaching.

Claude Bernard (1813–1878) became the successor of Magendie, and even more famous as an experimenter in pathology, physiology, and anatomy. Originally a poet, he finally turned to medicine and science, and in 1869 became a member of the French Academy.

One of the results of the French fondness for pathological anatomy was an outgrowth, unfortunate in some respects, of specialism, which made its appearance early and spread to other countries, particularly to Germany, so that to-day there is scarcely an organ in the body which has not only its special student, but its special representative in medicine. It would be of interest to go over some of the various organs and count those who have become most renowned in the study of their diseases, but that is beyond the scope of this volume.

As Baas says, England, after her excessive participation in the iatrochemistry and iatromechanics of the seventeenth century, with a devotion that extended far into the eighteenth, seemed then to lose all confidence in systems and schools of medicine, inasmuch as since that time no system or so-called school has gained in Great Britain any large or permanent band of followers; even Brunonianism did not succeed in this respect. This form of conservatism is a characteristic of the British race. But while schools have not risen, individuals have formulated hypotheses or doctrines that at least attracted attention, if not followers. For instance, John Mason Goode (1764–1827) formulated an intricate nosological arrangement in his long-popular text-book entitled *The Study of Medicine*, and also arranged a classification of diseases now almost forgotten.

In 1816 Sir Charles Bell (1774–1842) made the memorable discovery that the posterior roots of the spinal nerves preside over sensation, and the anterior over motion; and this attracted anew the attention of English physicians to the nervous system, and was rewarded by the later discovery of reflex action or reflex phenomena, communicated to the Royal Society in 1863 by Marshall Hall. Both discoveries were important, and both were duly rewarded by yet others.

Benjamin Travers (1783–1858) seems to have been greatly influenced by the first of these discoveries, and led thereby to pay special attention to what he termed "constitutional irritation"; his studies on this subject are often quoted to-day, and are well worthy of perusal; he understood by this term a process (in strong contrast with inflammation) which subsides without hyperæmia and without plastic exudate, but which, on the other hand, may occasion liquid products and result in neoplasms.

Contemporaries of Travers were: Abram Colles; John Cheyne (1777–1830), of Dublin, who wrote on *Diseases of Children* and described "Cheyne-Stokes respiration"; William Stokes (1804–1878), also of Dublin, who distinguished himself in 1857 by a great work, entitled *A Treatise on the Diagnosis and Treatment of Diseases of the Chest*; Robert Graves (1797–1853), Professor of Medicine in the King's and Queen's College, Dublin, who published clinical lectures of his own, besides many clinical reports in connection with Stokes. Graves was one of the first to oppose the "absolute diet" of the earlier physicians in the management of febrile maladies, and requested that his epitaph should have but one line—"He fed fevers!"

"The School of Natural Philosophy" was the title applied to a system which, in Germany, ran parallel with that of Broussais, being the legitimate outcome of the medical philosophy of the eighteenth century which had originated there, and also a revival of opposition thereto on the side of realism. It led into speculative extremes, which finally sobered down, because of the meaningless scholastic phrases often introduced, and thus broke a path for the subsequent enthusiasm in behalf of French positivism in medicine. Those who constituted this school were, for the most part, men of importance, but were followed by a number of imbecile representatives. Use was made of the abstract doctrine of the philosophy of identity and the imponderables, such as electricity, mechanical forces, and magnetism, contrasted with which were the dimensions of matter and certain qualities, like sensibility, irritability, etc. Perhaps the greatest influence of this teaching was in the department of embryology and physiology, where Johannes Müller displayed his remarkable activity. Among the most distinguished representatives of the natural-philosophy school was Oken (1779–1851), of Bavaria, who subsequently taught in Munich, Jena, and Zürich, and published a large work on natural history, which did much for the popularization of this science; he explained that the skull is made up from a series of vertebræ; also discovered the Wolffian bodies, and was such a power in his way that Agassiz characterized him not only as "a master in the art of teaching," but as "a courageous and ruling spirit." Others of this school were: von Walther (1782–1849), eminent as a surgeon; Dollinger (1770–1841), of Bamberg, the distinguished leader of the Old Catholics; Reil and Prochaska, anatomists; Troxler and Schelling, philosophers and anatomists; Treviranus, the microscopist; Malfatti, Kilian, Spindler; Schmidt, of Vienna; and others too numerous to mention.

As a successor to the School of Natural Philosophy came the School of Natural History (1831–1850), which made important concessions to realism; its most prominent members were from South Germany. This school was based partially upon the philosophy of Nature, and expired almost suddenly. One of its most eminent exponents was Krukenberg, whose therapeutic creed was that " Physicians should be filled with pious reverence toward Nature; the organism is a whole, and must be contemplated in this sense; medical art is, undoubtedly, capable of decisive action, but let us not mistake that in many cases its activity is quite superfluous, in very many null and inadequate, and in many injurious." This school was the expression of the turn medicine was compelled to take in order to escape the after-effects of the one-sided, ideal, systematizing tendency of the eighteenth century (whose final outcome was natural philosophy), and to square itself with the realism and positivism of the nineteenth.

Schönlein (1763–1874), of Bamberg, outlined a system that taught pathological and anatomical revelations as concrete expressions of the independent entity disease, whose relation to the organism is as that of a parasite sojourning temporarily in it; he also constructed a classification of diseases, something after the manner of the botanical classification of de Candolle. One of his best-known pupils was Canstadt (1807–1850), whose *Jahresbericht* has preserved his name. Siebert, of Jena, famous as a diagnostician, and Haeser, the medical historian, belonged to this school.

An offshoot of the French school of pathological anatomy and diagnosis was the so-called New Vienna School, which aided the French system in obtaining high recognition in German medicine, and gained its first influence from the labors of Wünderlich (1815–1857); next to whom should be mentioned Baron von Rokitansky (1804–1878),—a Bohemian,—one of the most famous men in modern times, and who exercised a profound influence, even in foreign countries,—particularly in Italy and Russia. Von Rokitansky worked for a long time in miserable quarters in Vienna, but finally a magnificent building was specially erected for him. He was loaded with honors, and took his seat in the Austrian House of Deputies. Two sons are well known in medicine to-day, and two more have achieved reputation as singers,—a circumstance which the father embodied in the *bon mot* that "two of his sons howled and two of them healed." He transplanted into Vienna the tendency of the earliest pathologico-anatomical school, which captivated all by its novelty and interest, and in the post-mortem room and the clinical-lecture room he converted medicine in Germany to the realism of the nineteenth century. He was, indeed, the Van Swieten of his time in his influence upon educational affairs. His works are distinguished by simplicity, clearness, and logical order. He performed more than thirty thousand autopsies; for fourteen years he studied the defects of the septum of the heart and the comparative anatomy of the uterus and genito-urinary organs, yet paid little attention to the microscope or to applied medicine. He was a pathologist, pure and simple.

A friend and co-laborer,—Skoda (1803–1881),—also a Bohemian, was little, if any, less famous. In 1839 he gave to the world his famous work on *Auscultation and Percussion;* in 1847 became professor at Prague, and was the first man to lecture in German. In spite of his bachelor peculiarities, his taciturnity, and his heedlessness, he was very popular, and left a fortune,—quite in contrast to Rokitansky, who died poor. His scientific merit was based upon the fact that he overthrew the specific and pathognomonic arrangement of sounds, as taught by the French, and substituted therefor a category, based upon the physical constitution and shape of organs and tissues. He endeavored to develop a strictly scientific system of physics out of the empirical French doctrine of physical signs, and in his work on *Physical Diagnosis* he displayed an independent spirit, though as one who had received his impulse from France. He was the first in Germany to insist upon the merits of Avenbrugger, and was the leading diagnostician of his time of the new Vienna school. Skoda was the first for whom was created, in Vienna, a specialty after the French model,—that is, a special division for patients suffering from thoracic diseases. Great as he was, we must yet lay it up against him that through his influence,—first in Vienna and afterward throughout Germany,—practical medicine degenerated into simple diagnosis, and that, by his observations on the natural course of disease, undisturbed by therapeutics, he became the founder and exponent of expectant or nihilistic therapeutics,—the harbinger of a very cheerless period in the history of medicine.

CHAPTER X.

AGE OF TRANSITION (*concluded*).—*New Vienna School* (*concluded*): von Hebra, 1816-1880. Czermak and Türck, Jäger, Arlt, Gruber, Politzer.—*German School of Physiological Medicine*: Roser, 1817-1888.—*School of Rational Medicine*: Henle, 1809-1855.—*Pseudoparacelsism*: Rademacher, 1772-1849.—*Hydrotherapeutics*: Priessnitz, 1799-1852.—*Modern Vitalism*: Virchow.—*Seminalism*: Bouchut.—*Parasitism and the Germ-theory*: Davaine, 1811-1882. Pasteur, 1822-1895. Chauveau, 1827—. Klebs, 1834—. F. J. Cohn, 1828—. Koch, 1843—. Lister, 1827—.—*Advances in Physical Diagnosis*: Laënnec, 1781-1826. Piorry, 1794-1879.—SURGERY: Delpech, 1772-1832. Stromeyer, 1804-1876. Sims, 1813-1883. Bozeman, 1825—. McDowell, 1771-1830. Boyer, 1757-1853. Larrey, 1766-1842. Dupuytren, 1777-1835. Cloquet, 1790-1883. Civiale, 1792-1867. Vidal, 1803-1856. Velpeau, 1795-1868. Malgaigne, 1806-1865. Nélaton, 1807-1874. Sir Astley Cooper, 1748-1841. Brodie, 1783-1862. Guthrie, 1785-1856. Syme, 1799-1870. Simpson, 1811-1870. Langenbeck, 1810-1887. Billroth, 1819-1894.

A FEW of Skoda's more eminent colleagues deserve brief mention: Oppolzer (1808-1871) was singularly gifted in diagnosis, popular, a teacher of wide influence, and manifested in superlative degree the characteristics that constitute a great physician; he wrote little, but was for a long time Professor of Medicine at Prague. Von Hebra, the elder (1816-1880), worked a complete revolution in dermatology, and developed a classification based upon the pathological anatomy of the skin. He instituted a new and independent line of therapeutics as applied to this branch of our art, for which the medical world will ever hold him in grateful remembrance. Sigmund and Zeissel during the same period did much to clear up the problems of syphilis. To Czermak (1828-1873) and Türck (1807-1868) we are indebted (practically) for the making a specialty of diseases of the nose and larynx; of like service to ophthalmology were Jäger, Jaxtthal, Arlt, Stellwag von Carion, Hasner, Mauthner, Fuchs, and von Reuss, while Gruber and Politzer did as much for diseases of the ear.

An indirect offshoot of the new Vienna school is the so-called "Physiological Medicine," founded by Roser

(1817–1888), of Stuttgart (late Professor of Surgery in Marburg), seconded by Griesinger and Wünderlich. Their views were directed against the symptomatologists and idealists, and particularly against the School of Natural History, the claim being that physiology must include vital phenomena, while from the morbid portions of these phenomena the special science should be formed as an artificial, yet practical, division of knowledge. Wünderlich's book of therapeutics was for a long time the best guide in this direction, inasmuch as it left to individual thought and judgment—the Hippocratic method of investigation—the determination of value and demand. Another offshoot, that differs but little from this save in definition, is the "School of Rational Medicine," originated by Pfeufer (1806–1869) and Henle (1809–1855), and which, since 1841, has been represented by a special journal. While Wünderlich claimed pathology to be the physiology of the sick, Henle considered this questionable and made no distinction at all between the physiology of the healthy and that of the ill. The language of the followers of this school contrasted strongly with that emanating from other schools, and for a time was confident and ingeniously triumphant; nevertheless, it did not forget philosophical speculation, and Hegel may now be regarded as indirectly the godfather of rational medicine.

The vagaries of Paracelsus led indirectly, though positively, to the foundation of Homœopathy, and likewise originated the doctrine that bears the name of Rademacher (1772–1849). It is curious that this pseudoparacelsic system should spring up alongside of the Vienna school, its teachings being the classification of diseases by their therapy. Rademacher's followers possessed three universal remedies,—"cubic niter (nitrate of sodium), copper, and iron,"—and also three primary diseases that must take their titles from the three universal medicaments. In spite of the admission that these diseases were unknown, it was boldly asserted they were with certainty to be cured by the three chief remedies. The three primary diseases, "sodic nitrate, copper, and iron diseases," do not necessarily remain as such, as they may throw some organ "into a condition of sympathy, and thus it results that iron disease may express itself in the form of consumption, delirium tremens, etc., while a copper disease may appear as worms, paralysis, jaundice, etc." Besides universal diseases and universal remedies there were diseases of organs, to be diagnosed by the efficacy of organ remedies; thus, abdominal diseases must be relieved by corresponding "abdominal remedies," head diseases with "head remedies," chest diseases with "chest remedies," etc. Also for each particular viscus there must be a special remedy. What is the most surprising about this absurd doctrine is that it found followers, some even quite capable in their way.

Now, too, reappeared the Hydrotherapeutic System— the great apostle of which was Priessnitz (1799–1852)— based upon gross views of humoral pathology, according to which a disease entity was to be expelled in the form of sweat, eruption, etc. Poultices, cold packs, and cold baths were the principal therapeutic measures. Winternitz has made hydrotherapy popular and, in a measure, effective in the management of certain maladies.

Rudolph Virchow in 1858 instituted the doctrine or theory known as "Modern Vitalism," which, in fact, was borrowed from natural scientific medicine, and distinguished from the vitalism of the previous century in that it breaks up the old vital force, which was supposed to be either distributed throughout the entire body or located in a few organs, into an indefinite number of associate vital forces working harmoniously, and assigns to them all the final elementary principles without microscopic seat. "Every animal principle has a sum of vital unities, each of which bears all the characteristics of life. The characteristics and unity of life cannot be found in any determinate point of a higher organism,—*e.g.*, in the brain,—but only in the definite, ever-recurring arrangement of each element present; hence it results that the composition of a large body amounts to a kind of social arrangement, in which each one of the movements of individual existence is dependent upon the others, but in such a way that each element has a special activity of its own, and that each, although it receives the impulse to its own activity from other parts, still itself performs its own functions." This is nothing but another way of expressing the cell-doctrine, to which many medical men are now committed, which means that all bodies are built up of cells and that each cell has a unity and a purpose of its own.

In 1677 Sir Robert Hooke discovered plant-cells; later Schwann discovered animal cells and Robert Brown cell-nuclei; but it remained for Virchow to supply the gap which had risen between anatomical knowledge and medical theory; that is, to supply a "cellular pathology," since which time the cell has assumed the rôle which the fibre occupied in the theories of the seventeenth and eighteenth centuries. Time alone can decide as to the ultimate validity of this theory, which has in certain circles been most enthusiastically received. One of its weakest aspects is, perhaps, that the so-called intercellular substance plays an uncertain and unsatisfactory part. An important feature in which the cellular pathology differs from other systems, and particularly from the old humoral pathology, is in the doctrine that the blood itself is not the proper and original cause of dyscrasiæ, and probably not the cause of continuous alteration of the tissues; that these dyscrasiæ arise because the blood is not an independent structure, but dependent upon the condition of the patient in consequence of its continuous conveyance of the noxious material from all parts of the body,—the blood is, therefore, merely the medium for the production of the dyscrasia. This theory has made several peculiar, new, and symptomatic or morphological forms of disease, such as leukæmia, leucocytosis, etc. Virchow also cleared up the old and obscure ideas regarding pyæmia, and proved that an absorption of pus into the blood, which the name implies, is quite impossible; likewise, that pyæmia is inseparable from thrombotic processes.

Virchow was born in Pomerania in 1821, and in 1849

Fig. 37.—Rudolph Virchow.

distinguished himself by attaining the highest grade in the career of the learned,—a professorship, which he first held in Würzburg. During earlier years his residence and labors were largely the result of necessities arising from political views, for on account of these he was long denied a residence in Berlin. A personal friend, now old, once an *interne* in the great Julius Spital, in Würzburg, at the time when Virchow taught there, tells me a light was burning every night in Virchow's room until 3 A.M., and yet the professor was always out at work by 7. It was by such intense application that he arrived at his present position at the very top of the professional ladder; but very few men have the physique and constitution to stand such arduous study. In 1856 he assumed the chair of Pathological Anatomy in Berlin, and introduced microscopical anatomy, to which Rokitansky had not given sufficient attention. Virchow was a former pupil of Johannes Müller, famous as a physiologist and pathologist, and his views to-day are often tinged by the doctrines acquired from this great teacher. He is also a great admirer of Harvey, whose picture, at least for a long time, was the only one permitted to hang in his study. His first edition of *Cellular Pathology* appeared in 1858; the colossal work on *Tumors* in 1866, in which he carried out the division of morbid growths originally adopted by Johannes Müller in 1838, classifying them according to their microscopical elements. He is also scarcely more celebrated for his teachings and labors than for the number of famous pupils brought up under his influence, among whom may be mentioned Leyden, Recklinghausen, Cohnheim, Waldeyer, Kühne, and Rindfleisch. As a result of his labors has arisen in Germany what has been called the "Medical School of Natural Sciences," that seeks, by means of pathological anatomy and microscopy, experimental physiology and pathology, and the other applied methods, to make of medicine an exact science; and to it belong such men as Ziemssen, Gerhardt, Nothnagel, Liebermeister, Senator, Erb, Vogel, and others. An offshoot from this is the so-called "Munich Clinical School," to which belong von Buhl, Pettenkofer, Seitz, and Oertel.

The splitting up of medicine into specialties, and the increase of its subordinate branches into schools,—so called,—resulted in great danger to the unity of medical science.

A return to the methods which combine science and practice—the so-called clinical-practical method—is again sought by men who have established the well-known *Zeitschrift für Klinische Medicin*, under the management of Frerichs and Leyden,—a journal which has already done a great deal of good.

The versatile Bouchut, of Paris, has recently published a theory,—the so-called "Seminalism,"—for which the claim is made that he grants nothing to hypothesis, and everything to observation; its characteristic is that this new theory is also vitalistic,—in fact, the French have scarcely ever brought forth any other than vitalistic theories. Borden and Barthez, during the previous century, created the first French theory, which was followed out by Bichat, and later by Bouchut, who, as a matter of fact, owes much to Bichat. Bouchut teaches that beasts have an intelligence of instinct, and men one of abstraction; no beast oversteps the limits of animal thought, which is separated by an abyss from the productive thought of men; there is a proper kingdom of man, in accordance with his special nature; also, that the vital forces of men and of beasts are entirely different from each other, and that the principle of physical identity remains in the bodies of each, since the constantly renewed mass is formed in exact accordance with the original plans; in all the changes of his elements man is identical with man; all internal and external causes of disease modify, more or less, the vital force and its impressibility in the fluids or at some point in the economy, either increasing it or diminishing it. This theory, published in 1873, claimed "in the abstraction of its promise and completeness of its conclusions to yield to none of its predecessors." Yet, even in France, the task of transforming medicine into a natural and exact science is far from being a *fait accompli*.

The most recent theories of disease are the result of microscopical study of germs,—the germ-theory, in fact,—

and stand in the closest possible relation with the doctrine of spontaneous generation, fermentation, miasm, and contagion. In 1838 Ehrenberg regarded infusoria as animals, but Dujardin in 1841 expressed doubts, and Perty in 1852 affirmed that most forms classified as infusoria should be assigned to the vegetable kingdom, where, a little later, Naegeli relegated them. The correctness of this conclusion was proved by Cohn, who also perfected a classification. This particular form of investigation began in the twenties of the present century and assumed its present direction in the thirties and forties. Gaspard, in 1823, renewed the experiments of Haller, and injected into the veins of animals, not alone putrescent material, but the blood of other creatures suffering from the effect of such injections. Bassi, in 1835, discovered the cause of silk-worm disease, thereby giving special impulse to the theory of parasitism, and this was quickly followed by evidence of the existence of both vegetable and animal exciters of disease. Schoenlein, in 1839, demonstrated the fungus of favus; Vogel discovered the *Oïdium albicans* in 1840; Goodsir, the *Sarcina ventriculi*, in 1841; but the greatest influence upon the development of the parasitic, or germ-theory was the sequel to the discovery of the anthrax bacillus, by Davaine in 1850. In 1837 Latour and Schwann demonstrated that the cells, which were known even to Leeuwenhoeck, were actually vegetable forms, and Schulze had already pointed out that fermentation of fluids could only occur in the presence of extremely minute vegetable organisms; Chevreuil next showed that animal solids remained free from decomposition when protected from the access of germs; and in 1857 Pasteur demonstrated that fermentation and putrefaction were caused, not by chemical forms, as Liebig had taught, but simply by the agency of lower organisms, which he divided into aërobes and anaërobes; while in 1868 Chauveau queried as to whether morbific elements resided in the formed elements of germs or in their fluid constituents. Thus the theory of *contagium vivum*, for which Henle contended as early as 1821, was not forgotten. In Germany Klebs and Hueter became the prominent champions of this theory; Hallier had designated his so-called "*Microsporon septicum*," and introduced a method of fractional cultures. The views of Klebs were opposed by Billroth, who contended for his "phlogistic ferment" and "*Coccobacteria septica*," upon which he wrote an elaborate and extensively illustrated treatise; he also at that time opposed the specific character of the lower organisms as disease agents. Hallier's *microsporon* was refuted by Cohn, who studied and classified the various fungi, and distinguished between the pathogenic and the septicogenic,— that is, those which produced disease and those which produced ordinary putrefaction. Then came the experimental evidence of Davaine and Koch, who demonstrated the development of bacteria from spores. It is hardly necessary to discuss this theory further, but I may mention the labors of Panum and of Brieger, who deeply investigated the poisons produced by bacteria, to which are given the general titles of ptomaines and toxins.

It would be unjust, however, did I not mention the name of Lister in connection with the inestimable benefit that has accrued to surgery from the practical application of the theory of infection to wounds,—a measure that brought about an entire revolution in surgery and surgical technique, and an entire reversal of the statistics of operations; where thousands formerly died, thousands now live, their lives being indirectly due to the labors of this one man and his following.

I will add that it is necessary to realize the difference between life and death to appreciate the changes that have been brought about during the last score of years. Much that in former years was unjustifiable has become both justifiable and feasible; to-day patients, as a matter of course, live after operations which, so recently as when I was a student, were considered impossible, or if performed exposed the operator to the charge of manslaughter.

I have spoken of the impulse which came from Avenbrugger's invention of percussion, which was greatly extended through the translation of his work by Corvisart (1755–1821); the latter also excelled as a clinical teacher and pathological anatomist, and had much to do with the education of others of his *confrères* whose names are lustrous in history. Among the most celebrated was Laënnec (1781–1826), who, though brought up among most trying surroundings, early manifested a zeal for medicine. He became a field-surgeon in the French army soon after the Reign of Terror, and pushed his classical and medical studies with restless zeal. In 1815 his first experiments were made with the stethoscope, the invention of which was due to accident: in order to hear the sounds of the heart more clearly, he one day applied a cylindrical roll of paper, and then immediately constructed the whole form of the stethoscope upon the principle now everywhere resorted to. In 1819 he published his work on *Mediate Auscultation*,—a treatise on prognosis in disease of the lungs and heart, based principally upon this new aid to investigation. The treatise was speedily translated into all the languages of Europe. After enjoying a large practice Laënnec succumbed to ill health at the early age of forty-five. He seems to have had but slight appreciation of his own services to medicine, and to have prided himself rather on his skill in riding horseback. Honor and fame, however, followed closely upon the publication of his well-known work, and the manuals of physical diagnosis which now find frequent mention in book catalogues, and come from various and wide sources are the legitimate outcome of Avenbrugger's and of Laënnec's pioneer treatises.

A versatile French writer who devoted especial attention to medical nomenclature was Piorry (1794–1879), to whom we are indebted for the pleximeter. The double stethoscope, a legitimate extension of Laënnec's simple instrument, was invented by Cammann, of New York, and can justly be claimed for American medicine. Other methods of physical examination—like spirometry, chest measurement, and study of expired air—have been introduced since 1846. The ophthalmoscope, which has been of such sterling service, and is based upon the simplest of principles, was the invention of the famous Helmholtz, but just deceased. The principle of endoscopy,—the illumination and visual examination of the various cavities of the human body,—the various specula, the spectroscope, the sphygmograph, the more accurate record of physical sounds, the application of electricity, and the employment of thermometry represent a few of the strides in the medical science of the present century, thereby aiding and perfecting the art of diagnosis, which, in turn, must ever necessarily form the basis for all rational treatment. Let no one complain that we are still so far from certainty in every case; the wonder is that so much has been discovered in so short a space of time.

Wonderful as have been these advances, the greatest achievements have accrued to the department of surgery, which Chamisso terms "the *seeing* portion of the healing art." The sixteenth century opened the way for checking of hæmorrhages; the seventeenth accomplished great simplifications and improvement in the way of dressing wounds; the eighteenth gave a refining and elevating tendency to the study of applied practice, and raised surgery to a level with other branches of science; and now the nineteenth century has, toward its close, made surgery as nearly, perhaps, as it ever can be, an exact science, to which every other branch of science has been made contributory. The chain-saw, invented in 1806 by Jeffery, alone gave an impetus to resection, which was cultivated especially in Germany; to resection was added osteotomy by

Heine and Mayer; this, in turn, was succeeded by the so-called subcutaneous osteotomy of Langenbeck in 1854; Stromeyer introduced subcutaneous tenotomy in 1831, which was a very pronounced advance on all that had gone before; then came the introduction of anæsthesia, by which were made possible operations that had been beyond human endurance; by the introduction of the rubber bandage by Esmarch in 1873 bloodless methods were made possible. Pain and hæmorrhage, the two greatest enemies of the conscientious surgeon, being thus almost annihilated, there was left but an apparently theoretical limit to what the surgeon might accomplish. Orthopædic surgery, introduced by Delpech, was unknown prior to 1816; it was first practiced systematically by Stromeyer and popularized in France by Guerin. Operations on nerves were studied as special methods by Schuh, Wernher, and Nussbaum. Jobert and Simon abroad, and Sims and Emmet in the United States, by their studies of fistulæ peculiar to the genito-urinary tract in females, have conferred inestimable benefits upon suffering womanhood. So late as 1839 Vidal declared there did not exist in the history of surgery a single well-authenticated case of complete cure of vesico-vaginal tear.

It is not my intention to more than barely refer to the living surgeons of to-day, or those who have but very recently passed away from us; but it would be an injustice to overlook Bernhard von Langenbeck and Theodor Billroth. The former, born in 1810 and deceased in 1887, was for a time a teacher of physiology, but subsequently became successor of Dieffenbach in the University of Berlin. The influence he exerted upon surgery in Germany and (since the decline of French precedence) upon surgery all over the world, has perhaps been greater than that of any one man since Dupuytren's time. He it was that introduced into surgical technique and surgical pathology the experimental method of which Johannes Müller was the great exponent; indeed, the relatively high importance which pathology is given to-day in every surgical curriculum is due more to his labors than to those of any other one man. Genial, learned, indefatigable, he was the ideal accomplished teacher. It would be impossible in any short *résumé* of his life and labors to do justice to so distinguished a man, to whom the profession owes so much. Perhaps the highest testimonial that could be given would

FIG. 38.—BERNHARD VON LANGENBECK.
(From a photograph.)

be the enumeration of the men who were ever and always his enthusiastic admirers. Langenbeck was the founder of the German Congress of Surgeons, and for many years its president, and the permanent home this association has built for itself in Berlin bears his name; the surgical journal he founded has now passed its fiftieth volume, and is to-day the first periodical of its class in any country or language.

Theodor Billroth, who only within a few years joined the great silent majority, was for many years the surgical sun of Austro-Hungary, around whom revolved all the other lights of the profession in the empire. He was as expert with the microscope as with the knife, and equally great with both. Although his great and elaborate work on *Coccobacteria Septica* is now obsolete, it nevertheless marked an era in surgical pathology, as does also his text-

FIG. 39.—THEODOR BILLROTH.
(From an engraving of a recent photograph.)

book on the same subject, which reached fifteen editions and has been widely translated. He it was who made the first resection of the larynx and of the stomach, and to him we are indebted for many other daring operations. It was the fame of this teacher that in recent years led young Americans to Vienna, and he set the example in every way for a constantly growing number of students whose names are, or ere long will be, famous. Billroth was born, in

1819, in Bergen, and succeeded Schuh in Vienna, after having taught most acceptably at Zürich. What he was to his teacher, Langenbeck, such are the younger German surgeons, like Czerny, Gussenbauer, Mikulicz, and others, to *him*.

Here may be recalled the pride with which Americans greet the name of McDowell, who performed the first ovariotomy, and prepared the way for a branch of abdominal surgery the results of which have fairly astonished the world.

There is much to be said also for certain measures, such as the introduction into surgery of plaster of Paris, by Larrey; of starched bandages, by Scutin; of absorbable material for ligatures and sutures, the latter from animal sources. Finally, antiseptic—or, better, aseptic—methods of operating and caring for injuries and wounds have worked a revolution in methods and results that is, perhaps, the most important known to medical history.

At the beginning of the present century the French appeared to lead in matters surgical, and were distinguished by dexterity in operating, fertility of invention, accuracy of observation, and clearness of clinical teaching. The foundation of this reputation was laid by Desault, and upon it his successors continued to build. From his school descended the barber-surgeon Boyer (1757-1833), who became the first surgeon and trusted adviser of Napoleon, and was by him created a baron. He was the author of a work, in eleven volumes, which has survived many editions and translations, and therein he laid especial stress upon after-treatment. Richerand (1779-1840), like Boyer, was made a baron, and was a professor in Paris; but his character suffered from his overweening ambition and vanity; he was wont to exhibit most unpleasant personal traits; nevertheless his surgical ability entitles him to front rank among his contemporaries. The third surgeon honored with the rank of baron was Larrey (1766-1842), surgeon-in-chief to the Grand Army, and whom Napoleon I called the most virtuous of men. In 1792 he joined the Army of the Rhine, and was the physician of the so-called "flying ambulance" for twenty-two years. He was present in sixty great battles (including that of Waterloo) and four hundred engagements, and was three times wounded. His memoirs and monographs on subjects connected with military surgery cause him to be often quoted even at the present day. It is recorded that he performed two hundred amputations in a single day; during the march through Russia he had at one time in Smolensk ten thousand men to care for in one hospital. A wonderful organizer, he was idolized by the soldiers, and seems to have been held in nearly the same esteem as his great prototype, Ambroise Paré.

The most famous French surgeon of this century,— equally celebrated as a diagnostician, as an operator, and as a teacher,—was Baron Dupuytren (1777–1835). As a child he had been stolen, on account of his eminent beauty. His early life was one of poverty and want. He zealously devoted himself to anatomy and physiology as foundations for successful work in surgery, and ultimately secured a private practice that embraced all France, and, when visiting other countries, was received like a prince. For years he devoted three hours daily to didactic lectures. He died, leaving a fortune of several millions of francs; he even offered to the exiled Charles of England a million francs as a trifling recompense for his misfortune. He was known as "The Napoleon of Surgery,"— a title well earned, yet one which drew upon him the enmity of many of his contemporaries, particularly as he seemed inclined to persecute all who dared to tread in his path. His death resulted from empyema, for which he declined operation, preferring, as he said, "to die at the hands of God, rather than man."

The first truly scientific practitioner of orthopædic surgery in France was Delpech (1777–1832), of Toulouse, who was likewise the pioneer in subcutaneous tenotomy of the tendo Achillis and in autoplastic operations. At his own expense he erected a large orthopædic institute in Montpellier, and his death occurred while on his way to pay a visit to this institution, both he and his coachman being shot by an insane patient upon whom he had operated.

Dupuytren's successor in the Hôtel-Dieu was Roux (1780–1854), who earned specific reputation as a dextrous and rapid operator; his labors in constructive and plastic surgery were extraordinary. The first to apply physical investigation to surgery was Lisfranc (1790–1847),—best remembered, perhaps, in connection with amputation of the foot. Marjolin (1770–1850) was a teacher of eminence, as were also Sanson (1790–1841) and Cloquet (1790–1883), though the latter is better remembered for his works on anatomy than for his exploits in surgery. Civiale (1792–1867) is chiefly famous for revamping the operation of lithotrity, for, though a lithotrite had been invented by d'Etoilles, Civiale was the first actual operator, for which he was fiercely opposed by Larrey, Sanson, Velpeau, and others; he lived to see his rivals confounded and lithotrity accepted as a legitimate surgical procedure. Amussat (1796–1856) reinvented torsion of arteries for the repression of hæmorrhage, for, although this measure had been suggested by the ancients, it was held to be suitable only for very small vessels; he never held a professorship, yet at his residence were gathered so-called "conferences" that were attended by the most eminent medical men of the time; he is specially known in connection with the operation for opening the colon in the lumbar region. Pravaz was a surgeon of Lyons, whose name has been perpetuated by the small syringe—the original hypodermatic—which he devised. Vidal (1803–1856), of Cassis, made a reputation by his work on *Surgery*, in five volumes, which was extensively translated and reprinted throughout Europe. Jobert de Lamballe (1799–1867) rose from abject poverty to a professorship and a seat in the Académie; he is particularly remembered for his achievements in plastic operations. Velpeau (1795–1868) in 1834 became the successor of Boyer; popular as a teacher, and an author of great fertility, he devoted attention alike to surgery and midwifery; his *Operative Surgery*, in three volumes, and a treatise on *Diseases of the Mammary Gland* are still classics. Gerdy (1797–1856), like Velpeau, was the son of poverty-stricken parents; in 1833 he became a professor, and wrote extensively on bandages, dressings, and on operation for the radical cure of hernia. Bonnet (1802–1858) rendered great service to surgery by his researches upon diseases of the joints. Malgaigne (1806–1865), made Professor of Operative Surgery in Paris in 1865, devoted great attention to surgical anatomy, operative and experimental surgery, and especially to fractures and dislocations,—his work on fractures is met with on many book-shelves to-day. Nélaton (1807–1874) was surgeon to Emperor Napoleon III, and, though he wrote little, became peculiarly eminent as a practitioner; his ingenious probe, tipped with porcelain, by means of which he located a bullet in the foot of Garibaldi, is well known. He devoted special attention to tuberculosis of bones and joints, being, perhaps, further instigated thereto by the case of the Prince Imperial; his treatise on this subject forms most acceptable reading to-day, and he taught the existence of osseous tuberculosis long before such was recognized in either Great Britain, Germany, or the United States.

Were I to refer to living contemporaries of many of the celebrities just mentioned, I should speak with special reverence and esteem of Péan, Verneuil, and Ollier, who are to-day the greatest surgeons in France; but with their lives and labors any one may easily acquaint himself from sources which are at the command of all.

I pass now to the Italians, who, since Scarpa's time,

have never made any very decided impression upon surgery, although there are many most excellent practitioners of the art in Italy; the best known are Porta (1800–1875), Vanzetta (born in 1809), and Rizzoli (who died in 1880); Riberi, Tizzoni, Loreta, Durante, and others are, perhaps, equally entitled to mention.

Since the time of Gimbernat there have been no surgeons in Spain whose services have been sufficiently important to rouse special attention away from their native peninsula. The Spaniards are well educated, and well equipped for practice, but do not appear as great originators nor experimenters; doubtless because their medical schools and universities long since lost prestige, owing to clerical and Inquisitorial interference; nevertheless, Spanish medical literature has kept well abreast with that of other countries.

In Great Britain the example of John Hunter, during the eighteenth century, produced results of the greatest importance; advances were made simultaneously in physiology and pathology which the Anglo-Saxon races have been quick to utilize. While, perhaps, more conservative and less inventive than the French. the surgeons of England have ever been in the front rank, and quite early they gave great attention to careful local dietetic and hygienic measures, of which Continental surgeons were, and are, too often neglectful. English surgeons, too, while they have been specialists, have never been quite so narrow in their respective fields as the Continental surgeons, and it has always been rare to find one who was not also a good general practitioner; the immense advantages which this added knowledge confers must be apparent. The most celebrated representative of British surgery of this century was the son of a clergyman,—Sir Astley Cooper, born in Norfolk in 1748, but subsequently a resident of London. During youth he resolutely compressed the bleeding limb of a playmate who was the victim of an accident, so that time was gained for the arrival of a surgeon, who then tied the vessel; this decided his future calling, and he pursued his studies in London, Edinburgh, and on the Continent. In 1791 he settled down to private practice, which soon yielded him an income in excess of £20,000 ($100,000), for his day the equivalent of thrice that amount at present. At the age of seventy-three he succumbed to a long-standing asthma. He was a somewhat voluminous writer, and his works on fractures, dislocations, and diseases of the

FIG. 40.—SIR ASTLEY COOPER, BART.
(From a steel engraving of an original drawing by J. W. Rubidge.)

breast are by no means obsolete. His motto was: "First observe, and then think." Exceptionally endowed with all the graces of person, he became one of the most popular and influential men of his day; withal, he was always zealous for his profession, never unoccupied, and charitable to a high degree. Of his boldness we have evidence in the fact that in 1817 he tied the abdominal aorta, being the first to undertake this surgical feat.

A colleague of Cooper's at St. Thomas's Hospital was Travers, already spoken of in connection with irritation. Tyrrel, a nephew of Cooper, was a well-known surgeon, particularly in diseases of the eye. Others of the same name were: Samuel Cooper (1781–1848), who wrote a voluminous treatise on practical surgery; Bransby Cooper (1792–1853), Sir Astley's nephew and adopted son, who

FIG. 41.—SIR BENJAMIN COLLINS BRODIE, BART., F.R.S.

(From a steel engraving by J. Brain of a painting by H. Room.)

was well known, and who achieved an eminence that is only dimmed by that of his uncle.

Sir Benjamin Brodie (1783–1862) was distinguished as a special investigator and a soft-tissue operator, of whom it is said that, basing his actions upon his statistics, the older he became, the less frequently he operated. His best-known writings concern diseases of the joints. Guthrie (1785–1856), a man of noble characteristics, was the friend and companion of Wellington, whom he accompanied in all his campaigns. Although well known as a lithotomist, his fame rests chiefly upon studies and writings in the domain of military surgery. A colleague of his in the Westminster Hospital, Sir William Lawrence (1783–1867), was surgeon to the queen, highly esteemed as a dextrous operator, and an authority on ruptures and on operative surgery. John Lizars (1783–1861) was a pupil of John Bell, and distinguished himself as a bold operator and fertile writer; early in the century he treated chronic hydrocephalus by operation. Robert Liston (1794–1847) was another remarkable surgeon and a wonderful operator. Sir Charles Bell (1774–1842) has already been mentioned for his researches on the nerves, and he also wrote on operative surgery, and is somewhat famed for his opposition to venesection.

In Edinburgh James Syme (1799–1870) secured great reputation both by his dexterity as an operator—which is spoken of by his own pupils as marvelous—and by his introduction of resection into general practice. Sir James Y. Simpson (1811–1870) aided to make the Edinburgh school famous by his researches into the domain of both surgery and obstetrics. Though the inventor of acupressure, his name will forever be associated with the introduction of chloroform. Professor Dunn says that, "after seeing the terrible agony of a poor Highland woman under amputation of the breast, Simpson left the class-room and went straight to Parliament House to seek work as a solicitor's clerk. But on second thought he returned to the study of medicine, asking: 'Can anything be done to make operations less painful?' The ultimate result was the discovery of chloroform, and so the suffering of one became the occasion of the deliverance of many." Upon his advocacy of chloroform in obstetrics he had to defend himself against most vehement attacks of both Scotch and English clergymen, who affected to regard such procedure as a crime that transgressed the will of the Deity; but he successfully confounded these assailants with their own weapons, proving himself their more than equal in knowledge of Scripture lore.

Many other British surgeons, living and dead, deserve most honorable mention, but time and space will not permit. I cannot, however, pass by without mentioning Curling, Annandale, Chiene, Cheyne, Macewen, Ogston, Jonathan Hutchinson, Sir James Paget, Christopher Heath, Thomas Langmore, Savory, Holden, Holmes, Adams, Sir Joseph Lister, and Sir Prescott Hewitt, of the value of whose labors I have already tried to speak; Sir William Ferguson, of whom it is said that he had the eagle's eye, a lion's heart, and a lady's hand; John Bowman, best known for his work in ophthalmic surgery; Sir Henry Thompson, the eminent lithotomist and lithotritist; and Sir Spencer Wells, Keith, Lawson Tait, and Bantock, whose names are inseparable from the history of abdominal surgery. And what can be said of the young men who are being trained in the methods and practice of their predecessors—trained not only in the direction of manual dexterity, but in experimental science, to which they make the former subservient? All honor to these scions of Great Britain's surgical art, who have astonished the world with their consummate ability! I would that time permitted recapitulation of the work accomplished in late years by the present generation of men in London, Edinburgh, and other medical centres; but the scope of these chapters does not cover this ground.

CHAPTER XI.

HISTORY OF MEDICINE IN AMERICA.

The Colonial Physicians. Medical Study under Preceptors. Inoculation against Small-pox. Military Surgery during the Revolutionary War. Earliest Medical Teaching and Teachers in this Country. The First Medical Schools. Benjamin Rush, 1745-1813. The First Medical Journals. Brief List of the Best-Known American Physicians and Surgeons.

THE history of medicine in America commences with the early struggles of the physicians in the American colonies. One Dr. Wootton came to Virginia in 1607 as Surgeon-General of the London Company. The following year Dr. Russell was with Captain Smith in his exploration of Chesapeake Bay. Neither of these men stayed long in the country, since, in 1609, Captain Smith, after being wounded, was compelled to return to England for treatment, for lack of medical aid.

When, in 1626, Peter Minuit *purchased the island of Manhattan for the sum of twenty-four dollars*, there was probably no physician there at the time. Undoubtedly the first physician, in what is now New York, was Lamontagne,—a Huguenot, who arrived in 1637, and who seems to have been a man of great capability for his time. It would appear that men of no little eminence left the Old World for the New during the early days of the American colonies, and that the medical services which the early colonists received were on a par with those received by those whom they left behind in their old homes. During the seventeenth century a number of reputable physicians emigrated to this country, among them Dr. John Clark, of Boston, in 1638, whose son and grandson followed him in his profession and became prominent in their chosen calling. In 1644 came Dr. Child, a graduate of Padua, who seems to have been a man of great learning.

A number of younger Americans also went abroad

to study,—Leyden, Paris, Padua, and the British universities being those most eagerly sought. In Virginia, so early as 1619, the Colonial Assembly discussed the erection of a university or college. In 1637 a public college was established in Cambridge, and in 1638 the Rev. John Harvard left to it his library and half his fortune, after which it was called Harvard College. William and Mary College was chartered in Virginia in 1693. Probably the first lectures in anatomy given in this country were those of Giles Firman, which were given previous to 1647 at Harvard College.

It was in this early day that arose the custom, continued until very recently, of studying medicine with a preceptor. This was necessary at that time, and until comparatively recently, because of the scarcity of institutions of learning and the expense connected with an education. The form of apprenticeship was often gone through with for a term of years varying from three to seven, during which time the young student performed the most menial duties, had very meagre opportunity for anatomical study, and acquired his knowledge rather by contact with and absorption from his preceptor than in any other way. In this method of teaching the personal element was so pronounced that everything, in fact, depended upon the preceptor, save what natural talent and industry might accomplish. With such meagre opportunities the means for doing were equally scant. Nevertheless, emergency made many of these early American practitioners self-reliant and competent to treat, according to the knowledge of that day, the various accidents then so common. In 1636 the Assembly of Virginia passed a fee-bill for surgeons and apothecaries, fees, however, being often paid in tobacco, powder, lead, wampum, etc. Not a few combined ministry of the body and the soul, and a number of eminent physicians were also preachers of more or less renown,—among them John Rogers, John Fisk, and others.

Probably the only medical work published in America during the seventeenth century was *A Brief Rule to Guide the Common People of New England how to Treat Themselves and Others in the Small-pocks or Measels*. This was printed and sold in 1677, by John Foster, of Boston. It was printed upon one side of a single sheet in double columns, and described both of these diseases as due to the blood endeavoring to recover a new form and state.

The old English distinction between physician and surgeon was for many years quite generally preserved, but could not persist, because of the different conditions under which men practiced. During this century, also, a number of midwives made excellent practitioners,—among them the wife of Dr. Fuller, one of the *May Flower* pilgrims. Those colonial days, however, seem to have been free from the ravages of itinerant specialists and charlatans, who so abundantly infested Europe at the time. It is also to the everlasting credit of the American profession that it took no part in the horrible delusions and scandalous transactions connected with the Salem witchcraft.

By the beginning of the eighteenth century the population of the United States was about three hundred thousand whites ; by the end of the century it had increased to a total of about four millions. During this century a larger proportion of educated medical men came from abroad and settled in various parts of the country, while the Colonial and the Revolutionary Wars offered ample opportunity for the development and study of military medicine and surgery. Commerce between the two continents increased ; communication became more free, and the people of the Old World and the New were constantly brought into closer relation. The most lively medical controversy of the century was, probably, that excited over the introduction of vaccination against small-pox. In previous sketches I have had to intimate that the greatest enemy of the medical profession in time past has been the clergy. In this particular instance, however, it was to the Rev. Cotton Mather, of Boston, that the profession is largely indebted for the favor with which the new method was received in this country. In 1721 he called the attention of various American physicians to the method, then in vogue in Turkey, of inoculation with virus from the active disease. Dr. Boylston, of Brookline, Mass., who settled in Boston, corresponded with members of the British Royal Society and finally determined to put the method to actual proof. In 1721 he inoculated his own son with the virus of natural small-pox, and within the next year had inoculated two hundred and forty-seven persons, of whom about two per cent. died of the disease; while, of nearly six thousand persons attacked by the disease in the natural way, more than fourteen per cent. died. In spite of this, the man and the method were violently attacked by the people and the profession, and found their warmest defenders among the ranks of the clergy. Benjamin Franklin, then only sixteen years of age, joined with the rabble in opposing the inoculation method. Boylston was threatened with hanging, and had even to hide himself for a time. He died in 1766.

After the great discovery of Edward Jenner societies were formed for the promotion of vaccination all over the world. The earliest vaccination in the United States was performed by Dr. Waterhouse (born 1754, died 1846), who operated upon four of his own children in 1800.

It was during the eighteenth century that a number of our best-known educational institutions were founded in the different colonies,—among them, Yale College, in 1701; Princeton (College of New Jersey), in 1746; University of Pennsylvania, in 1749; Columbia (King's College), in 1754; and others only a little less known. In most of these latter were established medical departments, but the method of apprenticing students to physicians was still in general observance, no preliminary education whatever, as a rule, being demanded. In 1766, however, the New Jersey Medical Society ordained that no student be taken as an apprentice by any member of the society unless he had competent knowledge of Latin and some initiation in the Greek. About the middle of the century Drs. Bard and Middleton, in New York, and Dr. Cadwallader, in Philadelphia, began giving lectures in anatomy, while at Newport, Rhode Island, Dr. William Hunter, between 1754 and 1756,—a near relative of the

FIG. 42.—B. WATERHOUSE, M.D.

famous Hunters of London, and a pupil of the elder Monro,—gave a course of lectures on human and comparative anatomy. Dr. William Shippen, Jr. (1736–1808), —a student of John Hunter's,—returned in 1762 to America, and gave his first course of lectures on anatomy and midwifery during the years immediately following. His lectures led to the formation of a Medical Department of the College of Philadelphia, in 1765, in which lectures were continued regularly until the winter of 1775, when the War of the Revolution interfered. In July of 1776 Shippen was made Chief-Physician of the Continental Army, and in the following year was elected by the Provincial Congress Director-in-General of army hospitals. During the latter years of the war he returned to Philadelphia each winter, and delivered a course of lectures,

FIG. 43.—SURGEON'S HALL, NEXT TO PHILADELPHIA DISPENSARY, FIFTH STREET BELOW LIBRARY STREET.
Foundation of the Medical Department of the University of Pennsylvania. First medical school in America, 1765–1807.

shortened by the necessities of the case. Thus he was the first public teacher of midwifery in this country. He was ably seconded in his work by Dr. John Morgan (1735–1789),—also a pupil of Hunter and Monro, who received a prominent army appointment in 1775, but who, two years later, was unfortunately dismissed on charges subsequently proved false. Shippen and Morgan were for some time the only professors in the Medical Department of the College of Philadelphia. In 1768 Kuhn—a pupil of Linnæus—was made Professor of Materia Medica and Botany; and Benjamin Rush, a year later, was given the Chair of Chemistry. The commencement of this institution occurred in 1768, when the degree of M.B. was given to seventeen graduates. In 1779 political reasons led to the abolition of the College of Philadelphia, the University of Pennsylvania taking its place. Ten years later the former institution was restored, and in 1791 the two institutions were united. The present Medical Department of the University of Pennsylvania is, therefore, the legitimate continuation of the first medical school in America.

The Medical Department of King's College, New York, now Columbia College, was organized in 1767, by Clossey, an Irishman; Middleton, a Scotchman; James Smith, a graduate of Leyden; Tenant, an alumnus of Princeton College; and Bard, who was by far the most eminent of the group, a Philadelphian by birth, who had studied under the best masters in England.

The Medical Department of Harvard University was organized in 1783. Most prominent in connection with it was Dr. John Warren, the first teacher of anatomy and surgery, and the founder of a family of eminent medical men, whose descendant, Dr. J. Collins Warren, is to-day an occupant of the chair of surgery in the same school. The Medical Department of Dartmouth College was organized in 1798 by Dr. Nathan Smith,—a man of great energy and unusual versatility.

While these medical colleges were developing their strength the medical profession were not idle, and institutions and libraries sprang up in various places. The Pennsylvania Hospital, for instance, founded in 1762, is to be credited with the oldest medical library in this country, many of its volumes having been selected especially for it by Louis, of Paris, and the famous Lettsom, of London. It now contains nearly fifteen thousand volumes. The library of the New York Hospital, not quite so large, was founded in 1776; that of the College of Physicians, in Philadelphia, in 1788. The profession of New Jersey organized the State Medical Society in 1765. In 1781 was founded the Massachusetts Medical Society. In 1787 arose the College of Physicians of Philadelphia.

In 1789 the profession of Maryland organized the so-called Medical and Chirurgical Faculty of Maryland, constituting thereby the same organization as the societies of other States. Before the close of the century, Delaware, New Hampshire, and South Carolina had also organized societies. In the larger cities extensive hospitals were also founded,—the Pennsylvania Hospital, in Philadelphia, in 1751, inside of which the first clinical instruction in this country was given by Dr. Thomas Bond. The New York Hospital began in 1769, simultaneously with the organization of the Medical Department of King's College. The first insane-asylum in America was built at Williamsburgh, Va., in 1773, although the charter of the Pennsylvania Hospital, dated 1751, provided for the care of lunatics, though not at that time in a separate institution.

The most conspicuous medical character of the century in American history was undoubtedly Benjamin Rush (1745-1813). He was one of Shippen's earliest students in anatomy, studied widely abroad, was a member of the Continental Congress, and one of the signers of the Declaration of Independence. After him is named Rush Medical College of Chicago. He was an extensive writer on a variety of subjects, not only professional, but political, philosophical, etc. He recognized but two kinds of remedies, —stimulants and depressants,—and held it to be the principal duty of the physician to decide as to which were most advisable in a given case. He called calomel the "Samson" of the materia medica, and his opponents contended that he was right, since it had undoubtedly slain its thousands. As an accurate observer of disease, he was correct and exact, and his descriptions are to-day both classic and reliable.

The study of practical anatomy has always been carried on in this country under great disadvantages. At first only the bodies of executed criminals were sparingly furnished.

FIG. 41.—BENJAMIN RUSH, M.D.
(From a steel engraving by R. W. Dodson of a painting by T. Sully.)

In 1788, in New York, occurred the celebrated "doctor's mob," which attested the vehemence of public objection to dissection, and which for two days defied the control of all the authorities. Secret dissections had been practiced in Harvard College so early as 1771, but the practice was against the law even for sixty years later in Massachusetts. Physiology, as such, was not taught in any medical school in this country during the century, and experimental physiology was practically unknown. Surgery was eagerly studied, especially during war times, and Dr. John Jones (1729–1791), of the King's College School, was, perhaps, the most eminent of the surgeons of his day. Others who vied with him were William Shippen, Jr., the first teacher of surgery in the College of Philadelphia; John Warren, of Boston; Richard Bayley, of Connecticut; Baynham, of Virginia; and McKnight, of New York.

The position of midwifery during the earlier years of the country may be, perhaps, understood by the following extract from the *New York Weekly Post-Boy*, of July, 1745:—

"Last night died, in the prime of life, to the almost universal regret and sorrow of this city, Mr. John du Puy, M.D., man-midwife," etc.

The first practitioner of obstetrics in New England was Dr. Lloyd (1723–1810), a pupil of Hunter and Smelley; while Dr. Shippen, in Philadelphia, endeavored to organize a school for the instruction of midwives, in which, however, he met with insuperable difficulties.

The first attempt to regulate practice in colonial times was an act passed by the General Assembly of 1760, providing for at least a form of examination in physic and surgery, registration, etc. The first medical journal to appear in the United States appeared about 1790. It was entitled *A Journal of the Practice of Medicine and Surgery and Pharmacy in the Military Hospitals of France*, consisting merely of translations from the French journals of military medicine. The first real American medical journal was the *Medical Repository*, begun in 1797 and discontinued in 1824.

The present century, now drawing to its close, saw in its earlier half the rise of a large number of American physicians and surgeons who have made their names illustrious for all time by their teachings, their writings, and their invention and originality. While it is, of course, invidious to select names, the following certainly deserve honorable mention in this list, without the slightest disrespect or intentional slight to many others whose names must be omitted for want of space.

John R. Cox (1773-1864), an early student of Benjamin Rush, filled the chair of Materia Medica and Pharmacy in the University of Pennsylvania, and published the *American Dispensatory* in 1806. Caspar Wistar (1761-

FIG. 45.—GEORGE B. WOOD, M.D.
(From an engraving by T. B. Welch, of Philadelphia, made from a daguerreotype by McClees and Germon, of Philadelphia.)

1818) was the author of a *System of Anatomy*,—held in great favor in his day as a text-book. Nathaniel Chapman (1780-1853) was Professor of Theory and Practice in the University of Pennsylvania until 1850. John Eberle held the similar chair of the Jefferson School from 1825-1831. The former wrote on *Materia Medica and Therapeutics*, the latter on the *Practice of Medicine*, both works being exceedingly popular. John W. Francis (1789-1861) taught obstetrics in the College of Physicians and Surgeons from 1826–1830. Franklin Bache (1792–1864) was one of the authors of the *Dispensatory of the United States of America*, published in conjunction with George B. Wood, who was Professor of Materia Medica in the University of Pennsylvania, and who wrote also extensively on his chosen subject in monographs and large works. Robley Dunglison (1789–1869) taught for a number of years in the University

FIG. 46.—ROBLEY DUNGLISON, M.D.
(From an engraving by A. H. Ritchie of a daguerreotype by M. P. Simons.)

of Virginia, but removed later to the Jefferson School in Philadelphia. He was a man of great industry and versatility, and wrote on a variety of subjects, his best-known work being his *Medical Dictionary*. W. E. Horner (1793–1853) taught anatomy and histology in the University of Pennsylvania, and will long be remembered for his researches in these branches. John W. Draper (1811–1882) made himself eminent as well by his researches in photography and in general science, as by the publication of his treatise on *Human Physiology*, which first appeared in 1853. Better known as physiologist was John C. Dalton (1825–1889), whose text-book is to-day studied in many colleges and who first introduced the method of vivisectional classroom demonstrations in our own school here in Buffalo.

FIG. 47.—AUSTIN FLINT, M.D.
(From a steel engraving by H. B. Hall.)

Alonzo Clark (1807–1887) was one of the most eminent teachers of medicine that this country has produced. Austin Flint (1812–1886) was also a famous teacher of medicine in New York, who made his first reputation in the then small school in Buffalo. His text-book on *Practice* is the most popular American work on the subject that has ever appeared, and is still in general use. William P. Dewees

(1768–1841) was the author of a treatise upon *Diseases of Children*, which reached a tenth edition and which rivaled the similar treatise of John Forsyth Meigs. The best-known teacher of dermatology and venereal diseases was Freeman J. Bumstead (1826–1879), author of the most popular work upon the latter subject that has been issued from the medical press. He was professor of these diseases at the College of Physicians in New York. His

FIG. 48.—ISAAC RAY, M.D.
(From a steel engraving by H. Meyer of an original painting in the British Museum.)

text-book vied with that produced by William H. Van Buren (1819–1883), who, in connection with Dr. Keyes (still living), wrote a treatise upon the *Surgical Diseases of the Genito-Urinary Organs*, including syphilis, which has been, since its appearance, exceedingly popular with the medical profession.

Among the best-known neurologists and alienists of the century since Benjamin Rush wrote his *Inquiries and*

Observations upon Diseases of the Mind (1812) was Dr. Isaac Ray, who, in 1838, published a work upon the medical jurisprudence of insanity. Dr. Brigham (1798–1849) was superintendent of the Utica Insane-asylum for some years before his death; and Dr. Kirkbride, who died in 1883, had been superintendent of the Philadelphia Asylum for over forty years. Dr. John P. Gray followed Brigham as superintendent of the Utica Asylum, where he remained for thirty-two years, and founded the *Journal of Insanity*.

The first independent writer upon diseases of the eye was Dr. Frick (1793–1870), of Maryland. As illustrating how little our present specialties were then separated, it is worth while to remark that Dr. Edward Delafield (1794–1875), who, in 1826, was Professor of Obstetrics and Diseases of Women and Children in the College of Physicians and Surgeons, New York, delivered at the same time a special course of lectures upon diseases of the eye. The first man in the United States to make these diseases his exclusive specialty was Dr. Williams (1822–1888), of Cincinnati.

It would be very wrong, in this connection, to omit the mention of the name of Oliver Wendell Holmes, the genial " Autocrat of the Breakfast Table," but recently dead at a ripe old age, who used to say that he was " seventy years young," who was for a long time Professor of Anatomy at Harvard Medical College, but who was much more widely known and endeared to the English-speaking public by his beautiful poems and most attractive prose writings,—who, as author of the *Chambered Nautilus*, for instance, will be remembered so long as the English language has a literature and is read. He rendered a great service to the medical profession by first calling attention to the contagiousness of puerperal fever. Of his prose writings, his medical essays—entitled *Currents and Counter-currents*—make perhaps the most delightful reading.

Not a few Americans deserve special mention as surgeons and surgical teachers of eminence during the past hundred years. Without being invidious, there must, nevertheless, be mentioned John Collins Warren (1778–1856), first Professor of Anatomy and Surgery in the Harvard School, under whose auspices ether was first administered for the purpose of surgical anæsthesia, and who was the founder, in 1828, of the *Boston Medical and Surgical*

FIG. 49.—PHILIP SYNG PHYSICK, M.D.
(From a steel engraving by R. W. Dodson of a painting by H. Inman.)

Journal. He wrote an extensive treatise upon tumors, and, it is stated, first successfully tapped the pericardium. Philip S. Physick (1768–1837), a pupil of Hunter, has been spoken of as the "father of American surgery," which he taught in the University of Pennsylvania. He was a tremendous worker, but wrote very little. He employed animal ligatures made of buckskin. John Syng Dorsey (1783–1818) was a nephew of Physick; taught anatomy in the University of Pennsylvania; wrote a treatise on surgery, which was the second surgical text-book published in this country, and was the first in the United States to tie the external iliac artery. He died at the age of thirty-five, at a time when he was giving promise of exceeding eminence. Nathan Smith taught in Dartmouth, Yale, and Bowdoin Colleges, and was considered the best man of his day in New England.

Ephraim McDowell (1772–1830) lived in Danville, Ky.

FIG. 50.—EPHRAIM McDOWELL, M.D.

To him is justly due the great honor of having performed the first rational and deliberate *ovariotomy*, which he did in 1809, his patient living for thirty-two years. The operation was performed without an anæsthetic, and considering the circumstances under which it was carried out has shed a lustre upon his name and brain which nothing can ever dim. By this performance he became practically the father of modern abdominal surgery, and to him Americans and Europeans alike are delighted to render all the honor that is his due.

Perhaps the most eminent surgeon of the country was Valentine Mott (1785–1865), a pupil of Cooper and Bell, who taught in the College of Physicians and Surgeons, New York, until 1840, and in the University Medical School until 1860. He was a man of exceeding boldness and brilliancy, whose operations were performed at a time when anæsthesia was unknown, or was in its infancy, and who probably did more work in the surgery of the vascular system than any other surgeon who has ever lived. He was the first to tie the arteria innominata,—in 1818. As Gross wrote of him, he had a record of one hundred and thirty-eight ligations of various large arteries,—a record probably never equaled. He was also the first to do a successful extirpation of the clavicle for tumor,—an operation which at that time was considered very formidable. Though not a great writer himself, he is best known among students as the translator and editor of Velpeau's large work upon operative surgery.

Dr. George McClellan (1796–1847) was the founder of the Jefferson Medical School, and its first Professor of Surgery. He was followed later by Dr. Thomas D. Mutter, who left his surgical museum to the College of Physicians of Philadelphia and endowed a lectureship there. J. K. Rodger, of New York; John Rhea Barton, of Philadelphia; William Gibson, of Philadelphia; Gurdon Buck, of New York; Willard Parker, of New York; Frank H. Hamilton, of New York, who made his reputation while teaching in our Buffalo school, author of a most popular and valuable treatise upon fractures and dislocations; and Henry B. Sands, of New York, were men of greatest prominence during the middle and latter portion of the present century, each of whom has contributed in his way either to the science or to the literature of surgery. The most prominent figure in American surgery of the past forty years was Samuel D. Gross, of Philadelphia, professor in the Jefferson School, to which he moved from Kentucky, where he laid the foundation for his attainments and reputation. He was an early writer upon surgical pathology and anatomy, but is best known for his elaborate

FIG. 51.—S. D. GROSS, M.D., LL.D.
(From a photograph.)

System of Surgery, in two large volumes, which has survived several editions and is still most highly esteemed. Among others who ought to be mentioned are Nathan R. Smith, of Baltimore, the inventor of the anterior splint; Paul F. Eve, of Nashville; John T. Hodgen, of St. Louis;

Daniel Brainard, of Chicago, and his successor, Moses Gunn; Alden March, of Albany; Henry J. Bigelow, of Boston, who performed the first excision of the hip in this country, in 1852, and who invented the method of crushing and removing stone from the bladder at a single operation, known as litholapaxy; and D. Hayes Agnew, of Philadelphia, who finished, before his death, a large and elaborate treatise on surgery, in three thick volumes.

Of obstetricians and gynæcologists America has had no lack, and, in fact, the United States may almost be said to be the first home of gynæcology. Dr. Bard was the first Professor of Midwifery in King's College, now Columbia, New York, and the author of the first work upon the subject published in this country. In Philadelphia, Dr. Thomas C. James (1756–1835) was the first distinct teacher of obstetrics, his chair falling later to Dewees, already mentioned, who wrote extensively on midwifery and the diseases of children and of women. The same chair in the University of Pennsylvania was filled later by Hugh L. Hodge (1796–1873), a man of great originality and independence, who published a most elaborate and beautiful work upon his branch, which will always remain a classic. Charles D. Meigs, professor in the Jefferson School, Philadelphia, was the first to direct attention to thrombosis as a cause of sudden death in childbirth. He wrote both on gynæcology and midwifery. Bedford, of Baltimore, was another popular teacher and writer, with whom deserves to be mentioned William H. Byford, of Chicago, who wrote on both obstetrics and gynæcology.

Gynæcology owes much to the efforts of American schools and practitioners. The first successful attempt of McDowell's, already alluded to, was imitated by Nathan Smith in 1821; and during the next forty years thirty-six ovariotomies had been performed by eighteen different surgeons, with a record of twenty-one recoveries. Probably the most prominent passed figure in American gynæcology is J. Marion Sims (1813–1883), born in the South, where he invented his well-known speculum in 1852, whose introduction marked an epoch in the treatment of the pelvic diseases of women. It was also in South Carolina, among poor negro patients, that he perfected his method of plastic operations in the vagina for the relief of

FIG. 52.—J MARION SIMS, M.D.

vesical fistulæ, which he later demonstrated in Paris to the astonishment of incredulous Parisian surgeons, who had almost uniformly failed in their attempts, and which he later successfully and brilliantly performed in all the capitals of Europe, where, as in this country, he enjoyed the greatest reputation. He was the founder of the great Woman's Hospital in New York, in 1855, an institution from which has proceeded more good gynæcological teaching than from any similar institution in the world. Other ovariotomists and gynæcologists of great merit were John L. Atlee, and his brother Washington Atlee, of Pennsylvania; Dunlap, of Springfield, Ohio; Peaslee, of New York, who wrote the first American treatise on ovarian

FIG. 53.—D. HAYES AGNEW, M.D., LL.D.
(From an oil painting of 1890 in the College of Physicians, Philadelphia.)

tumors; Kimball, of Lowell, Massachusetts; and D. H. Agnew, of Philadelphia, who is, perhaps, yet better known as a general surgeon because of his *magnum opus*,—his *Treatise on Surgery*, in three large volumes, already mentioned.

After this brief *résumé* of the names and achievements of the best-known American physicians and surgeons no longer living, it remains only to say a few words with regard to the general character of their work and attainments. It certainly was the case, during the earlier and middle portions of this century, that men had much to gain, beside addition to their vernacular, by study in foreign countries. Edinburgh and London were, at first, the centres to which men flocked; during the middle of the century they gathered in Paris, attracted by such men as Broussais, Velpeau, and others; after which the tide of travel turned toward Germany, where the government does more for the education of medical men and the furnishing of distinct opportunities than is done in any other part of the world. But, thanks to the influence of the foreign schools and the receptivity and natural quickness of the American mind, we have reached a point in this country when it is no longer necessary for American students to visit the foreign centres for this purpose, advantageous as these may be in many respects. The only feature in which we are yet lacking is the matter of government aid and the government control of medical institutions, by which better opportunities may be afforded for pathological study. Aside from this, and the centralization of cases which government control permits, it may be said that the Americans are in all respects as good practitioners as—and in most respects better than—their foreign colleagues. They evince more of humanity, more of real interest and care in their patients, and more consideration for their comfort and welfare; while, in all that pertains to fertility of invention, to originality of performance, and accuracy of work, they, as a rule, excel. Divested of glamour, American surgery, both general and special, is ahead of most of that which one can see abroad, and the therapeutics of the American profession certainly surpass those of any other nationality. No one need feel, then, that it is necessary to go abroad for any purpose, unless it may be that polish and wide range of general information that necessarily come from travel and observation among other nations and peoples. In practical medicine, then, as in practical living, America leads the world.

CHAPTER XII.

THE HISTORY OF ANÆSTHESIA.

Anæsthesia and Analgesia. Drugs Possessing Narcotic Properties in use since Prehistoric Times. Mandragora; Hemp; Hasheesh. Sulphuric Ether and the Men Concerned in its Introduction as an Anæsthetic—Long, Jackson, Wells, and Morton. Morton's First Public Demonstration of the Value of Ether. Morton Entitled to the Credit of its Introduction. Chloroform and Sir James Simpson. Cocaine and Karl Koller.

It is not, perhaps, generally understood that we owe the term *anæsthesia* and the adjective *anæsthetic* to the genius of Dr. Oliver Wendell Holmes, who suggested their use to Dr. Morton. The term *anæsthesia* is applied to the artificial loss or deprivation of all sensation, which may be either local or general. It should be distinguished from *analgesia*, which means simply freedom from pain, consciousness being retained. In this respect local anæsthesia is really local analgesia, although the terms are confused in this regard.

Anæsthesia, in its present sense, is truly a modern discovery, which is to be credited to the United States. In its less restricted sense, however, it is a condition brought about by numerous drugs,—intoxicants, narcotics, etc.,—some of which have been more or less in use for centuries. Anæsthesia is also a condition which may be produced in the hypnotic sleep,—a fact well recognized by the ancients, although the attention of scientific men was scarcely drawn to the fact until the days of the notorious Mesmer. The substances which may produce loss of consciousness may be taken intentionally or unintentionally, and may be taken into the stomach, beneath the skin, or, when gaseous, through the lungs, in which absorption of the same into the blood is very speedy. It is not at all unlikely that the curious effects ascribed to some of the ancient oracles were due to the inhalation of gases arising from natural springs or produced from other sources.

The most common source of narcotic drugs has always been the vegetable kingdom; and the peculiar effects of the juices or other ingredients of the poppy, henbane, deadly-nightshade, Indian hemp, mandragora, etc., have been sung in poetry, rehearsed in prose, and known from almost prehistoric time. Ulysses and his companions were stupefied by *nepenthe;* a draught of vinegar and myrrh, or gall, was offered to Christ upon the cross, as it often was to malefactors; and Herodotus speaks of a peculiar habit of the Scythians, who produced some stupefying vapor,—probably from the seed of the hemp. From Biblical times, at least, the most common narcotic seems to have been alcohol in some of its numerous combinations. Furthermore, the effect of hemlock has been celebrated since the days of Socrates, who was permitted to drink it in order to soothe himself during his last hour.

Mandragora seems to have had a great reputation in times past,—so much so that it is probable that more than one substance was included under this term. Apuleius, who lived about a century later than Pliny, wrote: "If any one is to have a member mutilated, burned, or sawed, let him drink half an ounce of mandragora with wine, and let him sleep till the member is cut away, without any pain or sensation." Among the Chinese and the Indians similar drugs seem to have been in frequent use, especially the *bhang*, ordinarily known as *hasheesh*. In many parts of the East something of this kind was administered to condemned criminals, as well as those compelled to undergo rude operations. It is said, also, that mild intoxication was produced among the fanatics of the East for the purpose of firing them to the point of heroic deeds, as it is also said that among the Druids the practice prevailed of partially stupefying the novitiates before initiating them into the most sacred and secret rites of their cult.

Guy de Chauliac was almost the only surgical writer of previous centuries who has referred to agents for the relief of pain, although during and before his time it was customary to give something to those about to undergo torture, by which to deaden their sensibility; and, though in the fables of all lands and all times something has always figured to which was ascribed the power of making people oblivious to pain or to the peculiarities of their situation, it is very difficult to learn just what, if any, particular composition was referred to or deserved such mention. There is allusion to something of the kind in *Romeo and Juliet*; again, in *Cymbeline*; and in one of Middleton's tragedies, published in 1567, entitled *Women Beware Women*, occurs this passage:—

> "I'll imitate the pities of old surgeons
> To this lost limb, who, ere they show their art,
> Cast one asleep, then cut the diseased part."

Larrey, in his military campaigns, noticed the effect of cold in diminishing sensitiveness, and suggested that cold might be made a useful local anæsthetic. Many surgeons used to operate upon patients under the influence of alcoholic narcotization. It was in 1776 that Mesmer arrived in Paris and became the exponent of so-called "animal magnetism,"—later termed "mesmerism,"—now known as hypnotism,—under the influence of which he reduced to the state of unconsciousness of pain (*i.e.*, analgesia, as well as the more complete condition, anæsthesia) a number of patients, who were operated upon without feeling the slightest suffering.

But, in spite of the earnest attempts of humane surgeons in various parts of the world, no agent had been discovered which was proven safe and generally effectual, up to the time, for instance, of Velpeau, who in 1839 wrote: "To escape pain in surgical operations is a chimera which we are not permitted to look for in our time."

The substance known as sulphuric ether has been

known since the thirteenth century, when, as it appears, Raymond Lulli made certain—perhaps ambiguous—references to it. In 1540 it was known as the sweet oil of vitriol. It was not called an ether until 1730, when Godfrey spoke of it as such. It was frequently referred to during the last century by various writers, and the first reference to its inhalation seems to have been published in 1795 by Pearson. In a work by Beddoes, on *Factitious Airs*, published at Bristol, in 1796, is a statement that "Ether in pectoral catarrh gives almost immediate relief, both to the oppression and pain in the chest." Beddoes also states that after inhaling two spoonfuls he soon fell asleep. Later it was in somewhat general use internally for mitigating the pains of colic. By 1812 it was often inhaled for experiment or diversion, its peculiar exhilarating effects being generally known. So it is, perhaps, not strange that so soon as it was definitely recommended for purposes of surgical anæsthesia, a number of claimants for the honor of its discovery should quickly arise.

It was the same with nitrous-oxide gas, which had been known for a number of years, and which was repeatedly used for the purpose of anæsthesia before the introduction of ether for the same purpose.

Chloroform was discovered in the year 1831 by Guthrie, of Sackett's Harbor, New York, and about the same time by Soubeiran, in France, and Liebig, in Germany. But, although before the profession for sixteen years, it was not recommended for the same purpose as sulphuric ether until 1847, and then by Doctor—later, Sir—James Simpson.

For all practical purposes we may limit further consideration of the history of anæsthesia to these three substances, and mainly to the consideration of the introduction and adoption of ether, which displaced nitrous oxide, preceded chloroform, and has held its own to the present day as the anæsthetic in most general use, although in many respects inferior to chloroform. But the glamour of history pertains mostly to ether, because of the peculiar difficulties and incidents attending its production.

For the honor of its discovery there are four claimants:—Crawford W. Long, of Danielsville, Ga.; Charles T. Jackson, of Plymouth, Mass.,—both physicians; Horace Wells, of Hartford, Vt., and William T. G. Morton, of Charleston, Mass.,—both dentists. It is only fair to each of these four men to consider briefly the merits of the claims made for each, while at the same time attributing the final success of the new agent to the happy accidents which permitted Morton to make a public demonstration of its power in the Massachusetts General Hospital, before such eminent men as Warren, Bigelow, and others, by whose influence and reputation the agent was at once received upon its merits. This was on the sixteenth of October, 1846,—a year which deserves to be memorable in the history of medicine.

Crawford Long graduated, in 1839, from the medical department of the University of Pennsylvania, and settled in Jefferson, Georgia, where it seems to have been a common thing to have what was known as "ether frolics," during which the exhilarating effects of the inhalation of the drug were matters of common sport and amusement at various small gatherings. Long himself frequently inhaled the drug and often felt its benumbing effects. It is stated that it finally occurred to him to give it a trial in a surgical operation, and that, in May of 1842, he removed a small tumor from the neck of a patient thus anæsthetized and without any pain. Owing to the sparseness of the population and the lack of dissemination of medical knowledge in those days, no public report was made of these operations, which produced nothing more than local town-talk. A young student of Long's, named Wilhite, kept a negro boy under the influence of ether for some time, to Long's surprise. Long lived one hundred and thirty miles from any railroad, and the first published account of his operations appeared in 1849, which was suggested by an account of Morton's work, which he had read in the editorials of the *Medical Examiner* for December, 1846. Long died in 1878, the unfortunate controversy in which the four claimants already mentioned participated being not yet concluded. Nevertheless, there is every reason to think that he is entitled to the credit of having first anæsthetized a patient with sulphuric ether for the purpose of producing insensibility to pain.

Horace Wells began the study of dentistry in 1834, in Boston, and later opened an office in Hartford, Connecticut. He seems to have been a young man of great ingenuity, continually making new instruments and devising new experiments. To him is to be credited the first operation ever performed without pain by the use of nitrous-oxide gas. In 1844 a Dr. Colton delivered a lecture in Hartford upon the subject of this gas. A young man who inhaled it, and became excited, ran against some furniture, badly bruising himself, but made no complaint of pain. Wells, noticing this, said to a by-stander that he believed that one, by inhaling a sufficient quantity, could have a tooth extracted or a leg amputated without pain. The following day he inhaled the gas himself and had a tooth extracted by a Dr. Riggs. Wells remained unconscious for a little while, and, on recovering consciousness, cried out: "A new era in tooth-pulling! It did not hurt me as much as the prick of a pin! It is the greatest discovery ever made!"

He at once began the manufacture and use of the gas, which became quite general in that locality. His attention was also called to the action of the vapor of ether, which Dr. Marcy, a physician of Hartford, suggested to him to try as a substitute for gas; but Wells, finding it more difficult to administer, discontinued it and confined himself to the use of nitrous oxide. A month later Dr. Marcy gave ether to a sailor for a small operation, the man feeling no pain. These experiences of Wells and Marcy occurred two years after Long's work with ether, each being in total ignorance of the experiments of the other.

In 1845 Wells visited Boston for the purpose of introducing nitrous oxide as an anæsthetic, and called upon his fellow-dentist and old partner, Morton, among others. He was discouraged, with his lack of success, returned to Hartford, and continued the frequent use of gas for a couple of years longer, but met with no encouragement in introducing it for general surgical purposes, on account of prejudice and fear upon the part of physicians and surgeons. Wells died in January, 1848, a few days before the Medical Society of Paris passed a resolution that to him is due all the honor of having first discovered and successfully applied the use of vapors or gases whereby surgical operations could be performed without pain. There stands to-day in Hartford the monument erected by the city and the State, with the following inscription:—

"Horace Wells, who discovered anæsthesia, November, 1844."

William T. G. Morton was born in 1819, and, after failing in business in Boston, in 1840 went to Baltimore and studied dentistry. In 1841 he entered the office of Horace Wells, above alluded to, as assistant, and in 1842 became his partner, after having introduced a new kind of solder for fixation of artificial teeth to gold plates. In 1843 this partnership was dissolved, Wells moving to Hartford, while Morton, in 1844, entered the office of Dr. C. P. Jackson as a medical student, matriculating in the Harvard School, but never graduating. After Wells's visit to Boston, during which he tried to introduce "laughing gas," Morton and he had numerous interviews, especially with regard to this gas. Morton was not well versed in chemistry, and sought the advice of his medical preceptor, Jackson, with regard to its manufacture. Asking why Morton wished to make it and being told the reason, Jackson suggested the use of sulphuric ether, just as Marcy had suggested its use to Wells, saying that it was easy to procure, safe in employment, and equally productive of results. He also stated that the students at Cambridge College often inhaled ether for amusement. On the evening of the same day, September 30, 1846, Morton admin-

FIG. 54.—WILLIAM T. G. MORTON, M.D.

istered ether for the extraction of a tooth, the patient stating that he had felt no pain. On the following day he visited the office of a well-known patent lawyer for the purpose of securing letters patent upon his supposed discovery. This lawyer, learning of Jackson's connection with the subject, took time to consider the matter, consulted with Jackson, and came to the conclusion that the patent must be a joint affair, neither one having exclusive right to claim it. But

Jackson, fearing the censure of the Massachusetts Medical Society should his name be connected with the patent, and Morton—as a dentist—having no such fine scruples, it was agreed that the patent should be made out in the names of both, but that Jackson was to at once assign his interest to Morton; in return for which he was to receive a ten per cent. commission. Meantime Morton called upon Warren, one of the surgeons in the Massachusetts General Hospital, who promised his co-operation and sent him an invitation to test his invention in the hospital on October 16, 1846. The clinic-room was filled when Morton placed the patient under the influence of his *letheon*, as he had named it; after which Warren removed a tumor from the neck of a young man, and, as it appeared, without pain. Upon the following day another operation was performed upon a young woman, with the same happy result, while on November 7th an amputation was made, entirely painlessly. At this time Morton endeavored to disguise the odor of the substance he was using by aromatic oils. It was not until the staff of the Massachusetts General Hospital declined to use an agent whose composition was kept secret that Morton revealed publicly the fact that this was nothing but sulphuric ether disguised by aromatics. From a report of the Commissioner of Patents, published a little later, the following paragraph is taken, the report being in the nature of a commentary upon the discovery:—

"It has been known for many years that the vapor of sulphuric ether, when freely inhaled, would intoxicate to the same extent as alcohol when taken into the stomach. . . . The fact has stood, further, upon the pages of science for many years that the inhalation of sulphuric ether was productive of temporary narcotic stimulant effects."

After the issuance of letters patent Morton began selling office-rights, such being the custom then, as now, among the dental profession, who are much more commer-

Fig. 5.—Dr. Morton making the First Public Demonstration of Etherization at the Massachusetts General Hospital Surrounded by the Medical Staff of that Institution, October 16, 1846.

cial in their proclivities than their brethren of the medical profession. The result was an almost endless litigation, with the development of the greatest personal animosity and rivalry between Jackson and Morton, as well as the friends and descendants of the other claimants. Morton wrecked his fortune and ruined his health in his efforts to get substantial recognition and remuneration from the United States Government; and the history of his repeated attempts to interest Congress and the various officials of the government, from the president down, is instructive, but far from pleasing, reading. In these attempts he practically failed, and died from an illness contracted through exposure, after maddening disappointment, although he had been the recipient of numerous honors and some small pecuniary recognition from societies and individuals. Morton died in 1868. In reviewing the history of his life and labors there is much to justify the inscription upon the monument erected to his memory at Mount Auburn Cemetery, Boston:—

"Inventor and revealer of anæsthetic inhalation, before whom in all time surgery was agony, and by whom pain in surgery was averted and annulled; since whom science has controlled pain."

Charles T. Jackson graduated at Harvard Medical College in 1829, after having led an already eventful career as geologist and mineralogist. He spent several years abroad, meeting many of the most distinguished men upon the Continent and displaying, in many ways, a great deal of scientific talent and mechanical ingenuity. In 1835 he opened, in Boston, the first laboratory for teaching analytical chemistry in the United States. A year later he was made State Geologist of Maine, and spent three years in this capacity. He also did a great deal of work upon the State geological surveys of Rhode Island, New Hampshire, and New York, while he was the first to call attention to the mineral resources of the southern shore of Lake Superior, where, in 1845, he opened up copper and iron mines. In 1846 and 1847 he became deeply interested in the subject and discovery of anæsthesia, and after the successful introduction of ether by Morton, in the Massachusetts General Hospital, set up the claim that it was he who had suggested it to Morton. In a pamphlet, published a little later, he states: "In the year 1837 I discovered that ether-vapor was superior to alcohol as a remedy for the strangling and toxic effects of chlorine-gas after inhalations for that purpose in my laboratory." He then relates how he administered the vapor to himself for the relief of the irritation produced by inhaling chlorine, and describes his sensations upon going to sleep and awakening. This claim in its entirety was a great surprise to both Morton and Wells, and led to a most unseemly discussion, which degenerated into a downright professional fight. After the death of Wells, Jackson and Morton both claimed that nitrous-oxide gas was not an anæsthetic, and that insensibility to pain could not be produced by it, in consequence of which the use of the gas was quite discontinued. It became, then, simply a question of priority as to the administration of ether for relief of pain during surgical operations. Wells being dead, this brought Long into the conflict. Jackson visited Europe again, and presented his claim before numerous societies in such a way as to be recognized abroad as the discoverer of anæsthesia. The relative merits of the whole controversy appear to have been pretty well summed up in a memorial sent to the Senate and House of Representatives by several hundred members of the Massachusetts Medical Society, which contains the following paragraph:—

"The undersigned hereby testify to your honorable bodies that, in their opinion, William T. G. Morton *first proved* to the world that ether would produce insensibility to the pain of surgical operations, and that it could be used with safety. In their opinion, his fellow-men owe a debt to him for this knowledge."

In the Public Garden in Boston there has been erected a monument to the memory of the discoverer of ether, the donor being, at the time, unable to mention the individual to whom it should be dedicated. Upon one face is this inscription:—

"To commemorate the discovery that the inhaling of ether causes insensibility to pain, first proven to the world at the Massachusetts General Hospital, in Boston, October, 1846."

Upon another face are these words:—

"In gratitude for the relief of human suffering by the inhaling of ether a citizen of Boston has erected this monument, A.D. 1867. The gift of Thomas Lee."

Morton's untimely death, largely due to disappointment and, as he thought, to persecution, has been already mentioned. In 1873 Jackson's mind became deranged, and he died in an asylum in 1880.

Sir James Paget has summed up the relative claims of our four contestants in an article entitled "Escape from Pain," published in the *Nineteenth Century* for December, 1879. He says: "While Long waited and Wells turned back and Jackson was thinking, and those to whom they had talked were neither acting nor thinking, Morton, the practical man, went to work and worked resolutely. He gave ether successfully in severe surgical operations, he loudly proclaimed his deeds, and he compelled mankind to hear him." As Dr. Morton's son, Dr. W. J. Morton, of New York, says, when writing of his father's claim: "Men used steam to propel boats before Fulton, electricity to convey messages before Morse, vaccine-virus to avert small-pox before Jenner, and ether to annul pain before Morton."

So much for ether. I have already stated that chloroform was discovered by Guthrie in 1831. But, though discovered in this country, it was first introduced as an anæsthetic agent in Scotland, by Simpson, who, in 1847, at the age of thirty-six, began to direct his attention to the discovery of some means of alleviating pain during childbirth, having a very large obstetric practice. Simpson was not satisfied with sulphuric ether, because of its strong and disagreeable odor, and inquired of his friend Waldie, Master of Apothecaries' Hall, of Liverpool, if he knew of nothing likely to be a satisfactory substitute. Waldie, acquainted with the chemical composition of chloric ether, suggested that chloroform be prepared from it and used. Simpson experimented with this in 1847, and established its anæsthetic properties, which he made known through a paper read on November 10th of the same year. It was arranged that upon the 13th of the month a public test should be made at the Royal Infirmary; but Simpson, who was to administer the chloroform, was unavoidably detained. Accordingly the operation was performed as of yore, without an anæsthetic, and during its performance the patient died upon the table. Had this death taken place during the employment of chloroform, it would have been the death-blow of that substance as an anæsthetic. The first public trial took place two days later, the test proving a great success. Simpson goes down in history, then, not as the discoverer of anæsthesia, but as the one who introduced chloroform for anæsthetic purposes. He died in 1870, and upon his bust in Westminster Abbey is this inscription:—

"To whose genius and benevolence the world owes the blessings derived from the use of chloroform for the relief of suffering."

It is a bit of most interesting medical history that after Simpson's announcement of his discovery he was violently and vehemently opposed by the Scottish clergy, who reviled him for endeavoring to relieve the pains of childbirth, basing their opposition upon the primeval curse: "In sorrow shalt thou bring forth children." And the beautiful ease with which Simpson refuted this childish sophistry must ever be memorable; for with one short argument he silenced his opponents and turned upon them the ridicule of the entire profession. For he reminded them that the first operation recorded in history was performed under anæsthesia, since, when God created Eve from one of Adam's ribs, he " caused a deep sleep to fall upon Adam."

Cocaine is now such a universally recognized local anæsthetic that there is the best of reason for referring to it here—the more so because it affords another opportunity to do honor to a discoverer who has rendered a most important service not only to our profession, but to the world in general.

The principal active constituent of coca-leaves was discovered about 1860 by Niemann, and called by him cocaine. It is an alkaloid which combines with various acids in the formation of salts. It has the quality of benumbing raw and mucous surfaces, for which purpose it was applied first in 1862 by Schroff and in 1868 by Moreno. In 1880 Van Anrap hinted that this property might some day be utilized. Karl Koller logically concluded from what was known about it that this anæsthetic property could be taken advantage of for work about the eye, and made a series of experiments upon the lower animals, by which he established its efficiency and made a brilliant discovery. He reported his experiments to the Congress of German Oculists, at Heidelberg, in 1884. News of this was transmitted with great rapidity, and within a few weeks the substance was used all over the world. Its use spread rapidly to other branches of surgery, and cocaine local anæsthesia became quickly an accomplished fact. More time was required to point out its disagreeable possibilities, its toxic properties, and the like, but it now has an assured and most important place among anæsthetic agents, and has been of the greatest use to probably ten per cent. of the civilized world. To Koller is entirely due the credit of establishing its remarkable properties.

The writer makes no apology here for having introduced two distinct chapters,—one upon the history of antiseptic surgery, the other upon the history of anæsthesia. First of all, they are the two grandest medical discoveries of all time; and, secondly, they are of Anglo-Saxon origin, —the one British, the other American. To the introduction of anæsthetics and antiseptics is due a complete revolution of earlier methods, complete reversal of mortuary statistics, and the complete relief of pain during surgical operations; in other words, to these two discoveries the human race owes more of the prolongation of life and relief of suffering than can ever be estimated or formulated in words. What an everlasting disgrace it is that, while to the great murderers of mankind, men like Napoleon in modern times and his counterparts in all times, the world ever does honor, erects imposing monuments and writes volumes of encomiums and flattering histories, the men to whom the world is so vastly more indebted for all that pertains to life and comfort are scarcely ever mentioned save in medical history, while the world at large is even ignorant of their names. For this reason, if for none other, these chapters find an appropriate place in a work of this character.

Those interested in a somewhat more elaborate presentation of this subject may find it in an anniversary address delivered by the writer on October 16, 1896 (the semicentennial of Morton's public demonstration), in the Medical School of the University of Buffalo, and published in the *Buffalo Medical Journal* of November, 1896.

CHAPTER XIII

THE HISTORY OF ANTISEPSIS.

Sepsis, Asepsis, and Antisepsis. The Germ-theory of Disease. Gay-Lussac's Researches. Schwann. Tyndall. Pasteur. Davaine. Lord Lister and his Epoch-making Revolution in Surgical Methods. Modifications of his Earlier Technique without Change in Underlying Principles, which Still Remain Unshaken. Changes Effected in Consequence. Comparison of Old and Modern Statistics.

MODERN surgery, and, in no small degree, modern treatment of all disease, have been so completely modified from previous methods by the introduction of the so-called antiseptic system that it seems to be only right to devote some time in such a work as this to a *résumé* of the history of the doctrines and experiments which have led to the perfection, as it would seem, of modern methods.

The adjective "septic" comes from the Greek word "sepsis," which is often transferred to the English, and which means "putrefaction," or that which is putrid, or undergoing decomposition. From this word are formed two others,—namely, "aseptic" and "antiseptic,"—the one implying the exclusion of all causes of putrefaction and complete freedom from it, the other referring to methods employed to antidote the effect or counteract the influence of the agencies which produce sepsis or destroy them while still within the living body. By general usage the term "antiseptic" has been construed as the more comprehensive; hence, the modern method is usually spoken of as "antiseptic surgery," and hence the title above: "The History of Antisepsis."

The principle underlying the resort to antiseptic methods is summed up in the expression, now so generally received, —the "germ-theory" of disease. It refers, in general, to the so-called zymotic, or infectious, diseases, whose manifestations are protean, which are all communicable by one means or another, but which are not all necessarily contagious; some of which, being not at all amenable to surgical treatment, are regarded as "medical" diseases, while others, which occur mostly in connection with surgical cases, or which lead to conditions requiring surgical relief, are usually spoken of as "surgical" diseases. As excellent and only too common examples of these zymotic diseases may be mentioned tetanus, erysipelas, puerperal fever, typhoid fever, and those varied conditions which are generally grouped under the term "blood poisoning." Those which most concern the surgeon, and those in which most remarkable relief has been obtained are erysipelas and the various forms of blood poisoning. These, in their varied manifestations, have, until recently, been literally the terror of surgeons, and in military hospitals, for instance, have been the cause of more deaths than have ever resulted from wounds directly upon the battle-field. In civil hospitals, as well as in general and private practice, the mortality from these diseases was, until twenty-five years ago, simply frightful; while frequently, and over wide areas of territory, endemics and epidemics of puerperal fever would result in the death of almost every lying-in woman. In consequence of this terrible death-rate surgeons were afraid to operate, and certain classes of operations, especially those on the abdomen and joints, were never performed, except under most exacting circumstances. But few of the present generation can actually realize the completeness of the changes brought about by the adoption of the germ-theory, and the practical effect of its use as a working basis for combating disease.

While no intelligent student at present denies that the infectious diseases—of which the above named are but a very few—are the result of the introduction into the body, from without, of minute living organisms, for the most part vegetable,—thus constituting them in reality, as they are often called, *parasitic* diseases,—but few are so familiar with the history of modern discovery as to appreciate the basis upon which it has been demonstrated. The proof of the germ origin of disease is the legitimate outcome of the discovery of the actual causes of fermentation and putrefaction.

Aside from the crude and often wild notions which have appeared here and there in literature of previous centuries, about the first accurate investigations bearing upon this subject were with reference to the cause of alcoholic fermentation. About the beginning of this century Appert published a monograph upon the *Art of Preserving Animal and Vegetable Substances*, which consisted in placing them in closely corked or stoppered bottles, and exposing these to the temperature of boiling water. Gay-Lussac, the celebrated chemist, noticed that so soon as these vessels were opened, particularly if much exposed to air, their contents began to at once ferment or putrefy. This led to investigations into the production of alcohol, and the antiseptic effect of pure oxygen-gas; from which he concluded that oxygen is necessary at the commencement of the process, but not throughout its continuance. Some thirty years later, Schwann, by the use of the microscope, then reasonably developed, discovered in fermenting substances numerous very minute globular bodies, which had the power of reproduction, and which were present in juices or fluids undergoing alcoholic fermentation, but not in others, and which he concluded to be the exciting cause. Schwann also discovered that if, in vessels sealed by Appert's method, he allowed air which had been previously heated to come in contact with the fluids, no change resulted; from which it was evident that it was something other than the gaseous elements of the air which provoked fermentation. Schwann's investigations were corroborated, in 1843, by Helmholtz.

Schwann's results were contested by Liebig, one of the most eminent chemists of his time, who proposed a very different theory, ascribing putrefaction to the absence of oxygen and to the upsetting of molecular arrangements. He believed that non-nitrogenous substances did not spontaneously undergo putrefaction when pure, but they must be brought into contact with some substance already undergoing change, which latter was called a *ferment*, and which converted the oxygen of the air into carbonic acid. According to him, the ferment was some material undergoing decomposition.

The next researches on this subject were those of Schroeder and Dusch, in 1854, who studied the question whether filtration of air would prevent the fermentation of boiled fluids to which such filtered air might have access. The material used for filtration was cotton-wool; and they showed that air filtered through it was deprived of the agencies which produce fermentation. Then came Pasteur, who repeated the experiments of his predecessors and elaborated and confirmed them. He also found that it was not necessary to filter the air of its contained particles, but that if it were simply left undisturbed until these had settled by gravity, it might then be brought in contact with putrescible substances without causing any putrefaction.

In 1870, in a lecture upon haze and dust, Tyndall demonstrated beautifully and in public the presence of countless particles in the air, as well as that these were the agencies operating to produce undesirable changes in organic substances. Both Pasteur and Tyndall, as well as others, showed, as did also Lister, that heat as well as filtration was sufficient to render these particles innocuous. As the result of these and numerous other experiments, by various observers, which there is no time here to recount, it was gradually and irrefutably established that the gases of the air, *per se*, are powerless to cause fermentation or putrefaction in boiled fluids or tissues, or in material germ-free when exposed. It was sufficient, in order to so purify the air, to either previously heat it or filter it through cotton-wool or through fluids inimical to germ-life, while the boiling of organic material or its subjection to the boiling heat of water was sufficient to destroy all germ-activity in it at the time, or, as we say now, to sterilize it.

In this way, and even before any minute and systematic study of bacteria,—*i.e.*, before the inauguration of bacteriology as a separate department of scientific study,—it was practically established that the agencies which produce putrefactive changes or fermentation were minute particles which were ever present in almost every substance, and that by heat or something corresponding to filtration it was possible to remove them or destroy their activity.

So much had been established without reference to the etiology of disease. In order now to study the germ-theory of disease as applied to man we must go back a little, neglecting the vagaries or the pure conjectures of the ancients, to the era of pure philosophic speculation,—perhaps to the days of Needham and Buffon,—to the middle of the previous century, when scientists and naturalists began to discuss the so-called spontaneous generation of life; for it is well known that fluids, like milk and others, abound with life after a few days of exposure; and it was supposed that the living organisms it contained had a spontaneous origin. This question of the spontaneous beginning of minute living forms was agitated for a century, or practically until Tyndall and Pasteur gave it its death-blow by their accurate and convincing demonstrations. There was no lack of experimentation, but there was lack of exact knowledge and of accurate deduction from facts observed. The bacteria—which at that time were usually spoken of as "monads" and "vibriones," because of their spontaneous motion—were found under varying circumstances, which, not being scientifically inquired into, led thinking men into a most perplexing condition of mind. The two most ardent recent advocates of spontaneous generation were Bastian, of England, who wrote an elaborate treatise upon the subject, and Jeffries Wyman, of Cambridge, Mass., who gave it the benefit of all his influence. But, under the influence of blows dealt from the side of the physical laboratory by Tyndall, and from that of the biologists by Pasteur, the theory was weakened and effectually killed, so that to-day no one thinks of such a thing. On the contrary, life seems to be inevitably the gift of a preceding organism; and while the real origin of life is as unknown to-day as ever, there is not a single fact in the possession of scientists now justifying the view that life can have a spontaneous origin. Moreover, the researches of Pasteur and others into alcoholic fermentation and the rôle played by the minute yeast-plant, and the early researches of Pasteur, Davaine, and Koch into the rôle of micro-organisms in producing disease in animals, and the scientific and elaborate study of bacteria and vegetable molds, inaugurated by Cohn and continued by many others, have as their legitimate outcome the creation of bacteriology as a science, and the establishment of the fact that the real condition in the so-called infectious diseases is one of fermentative or putrefactive alterations in the fluids and tissues of the living body, corresponding in minutiæ to the changes produced in saccharine fluids by the yeast-plant, or in decomposing animal or vegetable matter by the many known bacteria which are capable of producing such changes. To put it in another way, disease is simply an expression of the fact that these minute organisms, which are visible only under high powers of the microscope and which reproduce their kind with astonishing rapidity, gaining access to the surface or interior of the body, begin there to thrive and multiply, taking up from the living animal material for their own nourishment, thus robbing their host of that upon which his tissues must live, while at the same time, as the result of their activity, they produce various substances which, so far as they are concerned, are excretory in nature, and many of which are extremely poisonous to the animal organism which harbors them. Such a disease as puerperal fever, for instance, is simply an expression of the fact that within the living human body there is going on active putrefactive change, by which the internal cells are being destroyed, this destruction being progressive and often far-reaching; and that, as the result of their presence in the still living body, the noxious or toxic excretory materials of which they get rid are absorbed, in consequence of which such varying symptoms as nausea, fever, purging, vomiting, delirium, and many other symptoms are produced, the objective evidence of their local activity being the actual destruction of tissues, as is seen in cases of abscess, phlegmonous erysipelas, etc. The condition known everywhere as gangrene, when moist and offensive, is nothing but the putrefaction of tissues *en masse* which are not yet detached or separated from the living body of which they but recently formed a living part.

Experiments with organic material outside the body have amply demonstrated that such putrefactive processes can be checked by certain precautions,—such as filtration of air, heat, etc. It remained for the genius of Lister to show how similar processes of putrefaction and exclusion of germs could be made serviceable for the prevention of disease in the human race. To Lister, then, is due the credit of having originated the antiseptic system and brought about a condition long yearned for by surgeons throughout the world, but never previously attained. What a revolution he wrought by his masterly researches can be appreciated only when one compares the impunity with which surgeons now perform operations which, in the pre-antiseptic era were regarded as absolutely unjustifiable,—a conclusion amply warranted by the statistics of that era. Great as is the credit due to Lister, it is equally desirable to state that his work was, for the most part, based upon the researches of Tyndall, Pasteur, and Koch, which had established the germ nature of the terrible infectious diseases and the germicidal effect of filtration, of heat, and of certain other substances and methods which permitted of the development of his own system.

The antiseptic method, as it has since been known, was naturally at first crude, although its scientific basis has never been shaken; and that it has been since, in large

FIG. 56.—LORD LISTER, M.D., D.C.L., LL.D., P.R.S.
(From a photograph.)

measure, modified, and that surgeons now resort to little, if any, of the paraphernalia which first made it such a formal proceeding, in no regard shake the scientific nature of its foundation, but rather have tended ever to corroborate it and establish it more and more firmly. Lister began with the supposition that the air contains the germs which are most active and pernicious in producing disease. It has been since learned that air-contact is, perhaps, least of all to be dreaded. We, however, recognize the germs as always the efficient agents, though we have since learned that other sources of contamination are much more to be dreaded than air. It had been the custom, up to Lister's time, to observe usually the ordinary forms of cleanliness, but, not appreciating the multitude of germs which lurk in and about the skin, it had not been customary to scour and prepare it as we have learned to do since Lister's day. The ligatures and instruments which were used and the dressings which were applied, as well as the sponges used during the operation, usually went through the ordinary forms of cleansing; and yet Lister's investigations showed the utter inadequacy of such preparation. His most important object-lesson, however, was that everything that came in contact with fresh or bleeding tissues might carry infectious material (*i.e.*, germs), unless it had itself been thoroughly freed from their presence. Accordingly, the system taught the accurate preparation of everything,— from the skin of the patient, which was to be carefully cleansed and shaven, to the hands of the operator, which were to be scrupulously scrubbed, as well as those of every assistant who might handle or touch any of the instruments or dressing materials. It included, also, the careful preparation of sponges, sutures, and ligature materials, all of which were kept protected from air-contact and in antiseptic solutions until the moment of their use. The dressing materials were impregnated with substances like carbolic acid, which had been proven to be germicidal; and impermeable material, like oiled silk, was used to cover the surgical dressing, in order that fluids which might leak through should not come in contact with the air, which might permit of their putrefaction, while, at the same time, air from without could have no access to the deeper parts thus protected.

The original method of Lister was very elaborate, and included also the dissemination throughout the air of the operating-room of a vapor of carbolic acid, which was disagreeable, sometimes almost fatal, to operators and bystanders alike,—its use being based upon the notion that the air was the substance most to be dreaded. The instruments were placed in strong antiseptic solutions, usually carbolic, which were pungent and irritating to the hands of all that came in contact with them. So thoroughly and ubiquitously were antiseptic materials employed that it was soon learned that they were of themselves rather injurious to the best interests of the patients upon whom they were employed. Their use, of course, was contingent upon the notion, then everywhere prevalent, that powerful substances must be used in order to counteract the activity of the much-dreaded germs.

In the course of time, however, it was learned that the air was not so much to be dreaded as had been supposed, and that even if it came in contact with raw tissues infection did not certainly follow. It was found also that the antiseptic solutions which had been so freely used for irrigating or drenching the parts during an operation were by no means essential, and that tissues often healed better which had not been subjected to so much irritation. It was learned further that it was not necessary to impregnate dressings with these same solutions, providing, in the first place, they were carefully sterilized by the application of heat, which in time came to be used for the purpose of sterilizing everything not injuriously affected by it. In consequence, then, all dressing material, silk ligatures, instruments, nail-brushes, etc., were subjected to live steam or to boiling water for twenty minutes or more, which was demonstrated to be completely effective in the destruction of all organic or bacterial life. This, of itself, was a very great simplification of the antiseptic method. It was also demonstrated that the vital fluids of the animal body had of themselves great germicidal power, and that the strong antiseptic fluids previously used tended rather to impair this power than to enhance it. Accordingly, fluids for irrigation came to be used only when there was some noxious material to be washed away. It was found that fresh wounds healed most kindly when least irritated by applications of any kind, providing only that nothing came in contact with them which could infect them. And, in this way, as well as by resort to simpler rather than complicated procedures, there was gradually substituted for the so-called *anti*septic method that which is now everywhere recognized, and always practiced, when possible,—*i.e.*, the *a*septic method. This simply means that it is very much better to exclude germs than to permit of their access and then try to kill them after they have lodged. The aseptic method is, therefore, now in vogue, and among the best operators always the so-called *dry* method of operating, which means that, so far as possible, nothing not absolutely needed at the moment should come in contact with the field of operation. This has been, in many respects, a great advance over the older antiseptic method, though based upon absolutely the same recognition of causes, being only an improvement in technique.

The benefits of Lister's studies, and of that which has grown out of them, are simply incalculable. The surgical infections which, thirty years ago, were the dread of all operating surgeons, have practically disappeared from civil and military hospitals. I esteem myself fortunate in this,— that I have been a living witness of the benefit of change from the old to the new, since when I began my work, in 1876 (over twenty years ago), as a hospital *interne*, in one of the largest hospitals in this country, it happened that during my first winter's experience,—with but one or two exceptions,—every patient operated upon in that hospital, and that by men who were esteemed the peers of any one in their day, died of blood poisoning, while I myself nearly perished from the same disease. This was in an absolutely new building, where expenditure had been lavish; one whose walls were not reeking with germs, as is the case yet in many of the old and well-established institutions. With the introduction of the antiseptic method, during the two years following, this frightful mortality was reduced to the average of the day, and in the same institution to-day is done as good work as that seen anywhere. The same was true without exception in the great hospitals of the Old World; and in Paris, where, thirty years ago, famous surgeons would go from one end of the building to the other, handling one patient after another without ever washing their hands, and where erysipelas and contagion of various kinds were thoroughly distributed, as it were, impartially, now the successors of these very same men, employing modern methods, get results which challenge comparison.

The world has seen few extensive wars since the introduction of the antiseptic system; but, in such as have occurred, its incalculable value in military hospitals has been amply demonstrated. The modern soldier is now taught how to make a prompt occlusive and antiseptic dressing of the wound which he may receive upon the battle-field, which, from the moment of its attention, continues to be treated according to the same enlightened method after he reaches the field-hospital, or when sent to the rear; so that men now receive extensive injuries to joints and to viscera, which previously were either promptly fatal, or fatal within a few days from erysipelas and hospital gangrene, from which they recover with useful—often with nearly perfect—limb or function of part restored.

The military hospital of to-day is, therefore, robbed of the terrors which used to make it almost a charnel-house; hospital gangrene, the special dread of active army-surgeons in time past, has almost disappeared from the category of known diseases, and one of the greatest dangers menacing the modern soldier has been removed from modern civilized life. The method has met with universal adoption among all civilized races and peoples, and all this through the energy and talent of the originator, now Sir Joseph Lister.

With the recognition of the germ nature of so many acute diseases has come also systematic study of the use of antiseptics internally; and, while no such exceeding satisfaction has resulted from labors in this direction, we have, nevertheless, learned that most of the infectious diseases of the alimentary canal—for example, cholera, typhoid, etc.—are well attacked by means of antiseptics administered internally; that many of the conditions that depress and annoy are due to the presence of germs in the alimentary canal and the urinary system, and are best combated by means which shall remove these agencies, if not destroy them. It has been learned, also, that many forms of skin disease are parasitic, and that these are only successfully treated by the employment of antiseptics externally.

And so the recognition of the germ nature of infectious diseases and the germicidal properties of certain substances, now spoken of as antiseptics, have kept pace, the one with the other; and in consequence the world has reached a period in its medical history never even dreamed of by our forefathers, when the infectious diseases have been shown to be practically preventable and, to a large extent, curable by the employment of drugs directed especially against their exciting cause. What the years to come may have in the way of further discovery in this direction, we may not foresee. So far as one can at present see ahead, the next advances must be in the direction either of means which shall fortify the human organism against the inroads of bacteria, or disease-germs, or else in the discovery of substances, such as we do not yet know, which shall be at the same time poisonous to the germs and innocuous to the patient, to whom they may be administered in doses sufficient for their purpose. Any material possessing these properties would be an ideal antiseptic for internal purposes. At present we only approach our ideal, but are very far from its active realization. In no way would mankind be more greatly benefited than by the prosecution of studies which may lead to satisfactory results in either of these directions.

CHAPTER XIV.

AN EPITOME OF THE HISTORY OF DENTISTRY.

Rude Dentistry of Prehistoric Times. Early Instruments for Extraction Made of Lead. Dentistry on the Same Low Plane as Medicine during the First Half of the Christian Era. Dentistry Taught at the School of Salernum. Progress of the Art on the Continent. Prosthesis and Substitutes for Human Teeth. Introduction of Porcelain for Artificial Teeth; of Metal and of Vulcanized Rubber for Plates; of Plaster for Impressions. From being a Trade, Dentistry is now a Profession, in which Americans lead the World. Statistics.

The following is a *synopsis of an address* delivered at the opening of the session of the Dental Department of the University of Buffalo, in October, 1895. It is appended here because it is certainly apropos of the topics herein considered, the colloquial form being retained.

Called upon at short notice to welcome you here, and to offer remarks of general professional interest, it occurs to me to be retrospective for awhile and to consider the steps by which that which was once an exceedingly crude art has been developed until now it is an exact science. In other words, I would invite your attention, for a time, to the history of dentistry. At a time even before our combined art and science had a definite history we find that gold was used among the Egyptians for the purpose both of filling teeth and of supporting and directing them. In the bodies of many Egyptian mummies, especially of the higher class, there have been found teeth filled with gold or with wood which was covered with gold. It is known, also, that the Hindoos and Egyptians inserted artificial teeth and that some of these were made of wood, often covered with gold, and held in place by gold or silver bands and wires. Herodotus, who traveled so extensively in Egypt and wrote most entertainingly of his travels, has noted the division of medicine among the Egyptians into special branches and the existence of physicians, each of whom applied himself to one disease and not to more. "Some," said he, "are for the eyes, others for the head, *others for the teeth*, and others for internal disorders."

It is known, also, that about 300 B.C. Erasistratus deposited in the temple of the Delphian Apollo an *odontogogue*, or tooth-forceps, made of *lead*, intimating thereby that only those teeth should be drawn which were loose enough to be extracted with such an instrument.

Celsus, who was a contemporary of Christ and of Cæsar, was the first to recommend the use of a file within the mouth for the purpose of removing irritating edges and points of teeth. He also recommended bursting hollow teeth by putting into them pepper-corns, which should absorb moisture, swell, and thus break the teeth in pieces. He also recommended to take particular pains to try to shake or manipulate teeth loose before extracting them.

Galen, about 150 A.D., taught that teeth were true bones and that the canine teeth should be called "eye" teeth, because they were supplied by a branch of the optic nerve. Aëtius, 300 A.D., apparently discovered the foramina at the roots of the teeth through which the nerves enter.

In Rome false teeth and sets of teeth constructed of ivory and fastened with gold wire existed as early as the Laws of the XII Tables, and before the days of Roman civilization it is known that the Etruscans were skilled in manipulation of gold within the mouth, while your dean has described and has, I believe, in his possession beautiful examples of Etruscan work of this kind.

Among the Arabs, after the Arabian domination of the then civilized world, attention was paid to the teeth, although this was considered a very inferior part of the physician's work. Among these Arabians much later, and in spite of their study of Greek writers and their translations from the Greek, there may still be met such passages as this from Garriopontus, 1045 A.D.: "On the island of Delphi a painful molar tooth, which was extracted by an inexperienced physician, occasioned the death of a philosopher, for the marrow of the tooth, which originates from the brain, ran down into the lungs and killed that philosopher." For all that I know, this is the first record of a death after extraction of a tooth. Albucassis, 1100 A.D., gave directions for replacing lost teeth by natural or ivory substitutes. For centuries extraction of teeth had been and was considered a critical and dangerous operation, although itinerant quacks drew them without hesitation.

The Roman poets and satirists made many allusions, in their day, to the teeth and to operations performed upon them.

During the Middle Ages the most celebrated medical school that the world ever saw was founded at Salernum, and was for several centuries the headquarters to which resorted men who desired to study medicine and patients from all parts of the world who desired to be cured of various diseases. It was a favorite stopping-place for crusaders on their way to and from the Orient, and history relates many interesting episodes pertaining to such visits. Under the influence of this school dentistry was more or less cultivated by those who practiced surgery. Bruno, of Langoburo (about 1250), mentions various operations upon the teeth and the antrum, although that was nearly four hundred years before Highmore carefully described this cavity. Johannes Arculanus (Giovanni d'Arcoli), in the fifteenth century, filled teeth with gold. I must digress for a moment to speak of another suggestion of Arculanus's. You know that quite recently the use of the magnet has once more come into vogue among oculists for the removal of foreign particles of iron or steel from the anterior chamber or the globe of the eye. It was Arculanus who, some five hundred years ago, suggested extraction of iron splinters from the eye by means of the *attraction of amber electrified by friction.* (For School of Salernum see page 72.)

During the sixteenth and seventeenth centuries the French surgeons, especially Dionis and Verduc, made many practical contributions to dentistry. In 1728 Fauchard wrote in Paris the first complete work on dentistry, —*Le Chirurgien Dentiste, ou Traité des Dents.* Auzebi, of Lyons, wrote another. Le Cluse first mentioned the English turnkey for extraction. Jourdain introduced a number of new and appropriate instruments and new forms of artificial teeth. Bourdet, dentist to the king, made artificial palates. Porcelain teeth were first introduced in France in 1774.

Among the Germans cosmetic dentistry, though still the favorite field of charlatans, was greatly cultivated. Serré wrote a treatise on *Toothache in the Fair Sex During Pregnancy*, but the first public dental clinic in Germany was not established until 1855, by Professor Albrecht, and in Vienna. It has been in Vienna, among the Germans, that dentistry has been in time past most honored, and was taught when it was scarcely recognized in the other German universities. Private dental institutions were also first established in Vienna.

Of all the tooth-extracting instruments, the dental forceps in crude form is the earliest, the first on record, perhaps, being that deposited by Erasistratus in the Delphian temple, as already mentioned. For hundreds of years these instruments scarcely changed in shape. It was Garengeot who invented the *key*, early during the last century. Before that, and for awhile, dentists who had abandoned the forceps used an instrument known as *the pelican*,—said to much resemble the skid used by lumbermen.

Before artificial (porcelain) teeth came into use the following substitutes were employed, their estimated value being in accordance with the order in which I name them:

Human teeth, animal teeth, hippopotamus- tusk and teeth, elephant-ivory, and bone.

Human Teeth.—Transplantation of teeth was at one time very common. After being inserted, they were held in place by pivots and ligatures, springs, and upon bases. The pivot method also included the use of screws. Ligatures for fastening teeth were made of silk-worm gut,—which, now so common in surgery, was used for this purpose, perhaps, two hundred years ago,—of gold wire, etc. The method by ligatures is the earliest of all. Human teeth have always been more or less expensive if fresh, few people being willing to part with sound teeth except for a money consideration. In 1784 a Philadelphia dentist offered, in an advertisement, two guineas each for sound front teeth.

Animal Teeth.—These were largely used, being held in place the same way as above, the principal objection being that it was difficult, often impossible, to match human with animal teeth. It was found, also, that the latter decayed very much more easily.

Hippopotamus-ivory.—This was at one time very extensively used. It was carved into the shape of the missing teeth, and was held upon a base; or it was carved into shape as a base upon which to rest human teeth. Most often it was used as a base for pivoting. Not infrequently a block was carved out which represented gum, teeth, and all, and partial dentures of this complex type were often so deftly fashioned as to be very realistic, the part representing the gum being colored. Unfortunately no dye nor color in the mouth could be made permanent.

Elephant-ivory.—This was used for the cheaper grades of work, being less durable.

Bone.—Bone was still more objectionable, and was used for only the cheapest work.

Artificial porcelain teeth were first introduced in France in 1774 and in America in 1817. Those which were first made were so large, awkward, rough, and ill-fashioned, without attempt to represent the gum, as to bear no comparison to the artistic products of to-day. They were intended for the most part for attachment to ivory bases. The artificial dentures made for George Washington were of this general character, and, although they called forth his encomiums in a letter to his dentist expressing his gratitude, they would pass for very shabby productions to-day. One of the greatest advances in dentistry was the introduction of gold bases as a substitute for the baseplates previously made of ivory or bone. This is distinctly an American invention, and is to be credited to Gardette, of Philadelphia, who produced the first bases of this kind in 1787. Since then other metals have been used only because cheaper, none having the valuable properties of gold.

Gutta-percha was introduced for this and various dental purposes in England, in 1851, by Trueman. In 1851, too, came Goodyear's process of vulcanizing, which the dental profession were at first slow to avail themselves of, but which led, as its value was recognized later, to expensive and almost endless litigation.

Another most valuable American invention was that of taking impressions by the use of plaster. This was introduced about 1844–'45. This method permitted the making of socket-plates, which, of itself, was a long step in advance.

So much for a very brief epitome of some of the most interesting facts in the history of dentistry. Did time permit, the matter would warrant treatment at much greater length. But what now is to be said of the condition of dentistry to-day? First of all, that it is no longer relegated to charlatans and itinerants, but is studied, practiced, and honored by men of the ablest minds and of the highest type. There is to-day scarcely any branch of applied science which calls for greater qualifications or for greater combination of mental endowment and manual dexterity than does dentistry. We, in New York, find ourselves now in position where the State has assumed not only to regulate the practice of dentistry, but even to pass upon the qualifications of those who propose to study it. In the assumption of this task by the State there is paid, perhaps, the greatest possible compliment to its dignity and to its importance.

The great field of medicine is now altogether too vast, and the various branches which pertain to it are too complex, to permit a mastery of all its details by any one mind. The man does not live who to-day can be considered *facile princeps* in more than a few departments of medicine. Life is too short to permit of it, and the study is altogether too extensive. There is also a growing public demand for specialization of work, and there is probably more excuse for the perpetuation of dentistry as a specialty than for almost any other branch. Nevertheless, it is necessary constantly to repress a tendency toward a failure to comprehend the general principles underlying all medical specialties, and it has been hard, at least until recently, to impress upon the men of the dental profession that they were really only practicing a branch of medicine, and that, in disregarding a general and comprehensive knowledge of the fundamental branches, they were but poorly preparing themselves for the practice of a dignified specialty. Certainly dentistry makes as many demands for mechanical training, digital dexterity, familiarity with the properties of materials, etc., as does surgery, and in some respects even more. Of course, to a certain extent in these respects it is like a mechanical trade. The great trouble with the dental profession, until very recent times, is that they have regarded their work too much as a trade and not enough as a profession. By taking the latter view of it the work is ennobled and their interest for it cultivated. By taking the trade view of it they have lost those finer features which lift mechanical work out of the mere level of a trade. Moreover, men in time past have been guilty of altogether too much trades-union tactics, which are vehemently opposed to professional ethics, and this has been another feature to degrade rather than elevate dentistry.

This has been indeed a great misfortune, for men have been misled by the need for cultivation of their hands, or their manual powers, and have been persuaded away from a finer study of fundamental principles upon which the whole practice of dentistry should be based. And so it has happened that men have been so ambitious to become perfect operators that they have neglected anatomy, physiology, chemistry, and pathology, have even neglected odontology, sacrificing everything else to their work as mere artificers.

If one scrutinizes the subject properly, there is no reason why there should not grow up a class of men fitted to attend to any lesion of the mouth or of the parts adjoining. In other words, there is no reason why there is not more excuse for true oral surgeons than there is for any other class of specialists, save possibly those who treat the eye. Aural surgery, nasal surgery, pelvic surgery, rectal surgery, etc., are simply voluntary limitations and applications of general surgery to special parts; but he who attends to the teeth has to do so much work of a character which the surgeon is not called upon to perform in any other area, that I have always claimed the oral surgeon deserved a place, as he had a field, by himself. Nevertheless, the knowledge which shall fit a man for such work is not to be obtained in the ordinary dental course, nor in three years of study, even under the best of auspices. The man who would be an ideal oral surgeon must be not only generally familiar with anatomy and physiology, but must thoroughly know the embryology of the face and teeth, the physiology not alone of the organs of the mouth, but of all the secreting glands and the chemistry of all their secretions; not only the anatomy of the cranium, but general anatomy as well, and even comparative anatomy. He must be well informed in the explanations of all the congenital defects met about the face and mouth; he must be familiar not only with the ordinary principles of pathology and bacteriology, but he will find in the fluids about the mouth such a fertile opportunity for bacteriological study that, be he ever so expert or erudite, he has still much left to investigate in this direction. There is no disease-germ with which he can afford to be unfamiliar, and, as any form of tumor may be found in or about the mouth, he should be familiar with the entire subject of tumors and their surgical treatment.

Then, again, he must be familiar not only with the physical properties of metals and the various materials used in plastic dentistry, nor expert alone in the operations about the teeth, but, inasmuch as he has to cope with various wounds, injuries, and operations about the soft parts, he must be thoroughly familiar with the principles of wound-healing; with the causes of sepsis and the agents which produce it, and the means of avoiding it; in other words, he must have a general training in operative surgery, and, according to my ideal, which may be high, he should be a man able to do almost any operation in surgery before he limits himself to surgery of the mouth. Unless he have this ability, he will not do such operation as well as a general surgeon can, because the underlying principles are the same, and the latter will have the greater command over them.

When, then, this perhaps ideal man has become thoroughly familiar with the principles of surgical anatomy, operative surgery, surgical pathology, and bacteriology, in addition to the things already mentioned, then, and not until then, may he and should he assume to operate for harelip, cleft palate, cancer of the tongue, and various other lesions in the parts about the mouth.

I wish I could say and demonstrate more to impress upon you the important bearing of modern surgical pathology to dentistry. Perhaps I can give you no better illustrations than you can see in the studies and writings of Prof. W. D. Miller, of Berlin, of whom I am proud to say that he is an American, and that he is the only American occupying a professorship in a German university. In his studies on the causes of dental caries and upon the bacteria of the mouth he has identified and named nearly a hundred species of the bacteria, many of which he has shown to be the active causes of dental decay. He has done, then, for dental pathology in this direction what other eminent observers have done for the processes of suppuration and ulceration in other textures and tissues, and has helped to show that they are all evidences of pernicious germ activity. By his researches, also, upon inflammation in elephant-tusks, and the results of injury, mainly by bullet wounds, he has shown us that the phenomena attending these changes in dental tissues are practically identical with those in bone. His researches have done very much to explain the pathology of that common disease, pyorrhœa alveolaris, which is known to be but one expression of local infection, while the possibility of migration of infectious organisms and of metastatic lesions in other parts of the body, having their origin in infectious disease in or near the teeth, has been brilliantly demonstrated by his interpretation of well-known clinical facts.

That American dentists are most highly regarded abroad is more than a matter of every-day knowledge. It has got to be so now that a foreigner will purchase instruments of American make, and then advertise himself as an American dentist for the purpose of getting business, —a purpose in which, as a rule, he is quite successful. But let me stop here to do honor to another American dentist who is more highly honored abroad than one ever can be at home, and of whom it might be said, perhaps, that he has had more friends among the royalty and nobility of Europe than any other man of his time. This is Dr. Evans, who has lived for years in Paris, who was the personal friend of Napoleon III and the trusted guide and companion of the Empress Eugenie when she fled from Paris. While it may be said of him that the qualities that made him so universally popular were personal qualities, rather than professional knowledge, it must be said in reply that it was his eminent professional attainment which first brought him such influential friends.

But time presses, and I want, before closing, to say a little about dentistry in America. It was about 1835 that Dr. Harris, then residing in Baltimore, though born near Syracuse, conceived the modern idea of the scope and practice of dentistry. He was ambitious to put the dentists of his time upon a higher professional level, and to make of dentistry a specialty in medicine. He applied to various medical schools to found dental chairs, and to teach oral pathology along with dental mechanics, as one of the branches of medicine, the graduating degree to be M.D., as with other medical specialties. But the men of his time were so short-sighted and of such constricted mental calibre, and the dentists were so uneducated, that the Baltimore schools declined. He therefore established a separate school, being forced to take this step. This school was the Baltimore College of Dental Surgery, established in 1839,—the first in any country. The dentistry of that day was crude, and its teaching was comparatively inefficient. It was not until six years later that the next, the Cincinnati College of Dental Surgery was organized,—in 1845. Then, in time, followed Philadelphia. But all these colleges were separate institutions, teaching only those branches which it was held necessary that a *dentist* should know and having very little of medicine in their curriculum. They conferred the degree of D.D.S.

In 1868 Harvard University did what she ought to

have done at the outset. She opened a dental department and began the teaching of dentistry as a branch of medicine, establishing therefor a separate degree,—D.M.D.,—*Dentariæ Medicinæ Doctor*. In 1874 the University of Michigan established a dental department, and a little later the University of Pennsylvania did the same. These university schools gave an immensely widened scope to the study, which was made broader with each succeeding year.

There are now forty-five dental colleges in the United States. Forty of these are members of the National Association of Dental Faculties, organized for the purpose of securing uniformity in teaching and in graduating men. Membership in this association is a certificate of high standing and of comprehensive advantages.

Last year (1894) the number of students in dental colleges was 4979, while the number of graduates was 1208. At present nearly all the States have legislation governing the practice of dentistry, and often more strict than that regulating the practice of medicine. In New York the law places dentistry on precisely the same plane as medicine,—prescribes the same qualifications for matriculation, the same length of study, exactions for graduation, examination, etc. In other words, the law is quite as strict regarding admission to dental colleges as to medical. After 1897 at least a full high-school course will be demanded for matriculation, and from now on we may look forward to having a really educated dental profession.

INDEX.

Advances in other sciences, 153
Ægidius, 77
Æsclepiadæ, 11
Æsculapius, 7
Ætius, 49
Age of foundation, 12
Age of renovation, 12
Age of transition, 12
Agnew, D. Hayes, 295
Akenside, 213
Albinus, 164
Albucassis, 63
Alchemists and charlatans, the, 187
Alchemy, 141
Alexander of Tralles, 50
Alexandria, library of, 31
    school of, 32
Alkindus, 60
Ambulant physicians, 17
Amendment in medical affairs, 151
American teaching of to-day, 298
Amphitheatres, dissecting, 111
Amussat, 269
Anæsthesia, history of, 300
Anatomic period, 12, 30
Anatomy and physiology of Galen, 39
Anatomy, chairs of, 111
Andral, 245
Andry, 177
Anel, 215
Animalculists, 183
Animism, 196
Animists, 183
Antiseptics, history of, 317
Antyllus, 51
Arabic period, 12, 57
    review of the, 97
Archiaters, 53
Aretæus, 34
Argentier, John, 146
Aristotle, 28
Arlt, 253
Arnold de Villeneuve, 88
Asclepiades of Bythinia, 44
Aselli, 160
Astrology, 141
Astruc, 138, 214
Aubrey, 183
Auscultation, 262
Avenbrugger, Leopold, 210
Avenzoar, 64
Averroës, 64
Avicenna, 61

Bache, Franklin, 287

Bachtischua, 59
Bacon, Lord, 153
Bacon, Roger, 68
Baglivi, 162, 172
Baillie, 213, 224
Barba, 165
Barthez, 201
Bartholin, 161, 184
Barton, John Rhea, 293
Baseilhac, 214
Bayle, 245
Bell, Benjamin, 219
Bell, John, 219
Bell, Sir Charles, 219, 274
Bellini, 172
Benivieni, 114
Bernard, Claude, 247
Bernard the Provincial, 77
Bhang, 301
Bichat, 160, 162, 164, 208
Bienaise, 177
Bigelow, Henry J., 295
Bilguer, 215
Billroth, Theodor, 264
Blumenbach, 222
Boerhaave, 193
    influence of, 168
Boerhaave's, clinics, 167
    system of medicine, 194
    theory of inflammation, 164
Bonnet, 270
Borden, 201
Borelli, 160, 172
Borri, 176
Botal, Leonard, 146
Bouchut, 259
Boyer, 267
Boylston, Dr., 279
Bouillaud, 244
Boulot, 177
Bourgeois, Louise, 166
Braid, Dr. James, 204
Braidism, 204
Brainard, Daniel, 295
Brasdor, 214
Brigham, 290
Brisseau, 178
Brissot, practice of bleeding by, 118
British surgeons, modern, 275
Brodie, Sir Benjamin, 273
Broussais, 243
Brown, Dr. John, 205
Browne, Sir Thomas, 175
Brunner, 183
Brunonian doctrine, 205

(343)

Buck, Gurdon, 293
Bumstead, Freeman J., 289
Burking, 231

Cabalistic theory, 141
Cæsarean operation, re-establishment of, 134
Camper, Peter, 219
Cardan, Jerome, 142
Cardinal powder, 165
Carion, Stellwag von, 253
Casserius, 162
Cataract, first recognition of its seat, 178
Cathedral medical schools, 89
Cell, the term, 153
Cellular pathology, 256
Celsus, Cornelius, 36
Cesalpinus, 155
Chamberlain's invention of obstetric forceps, 166
Chapman, Nathaniel, 286
Charitable institutions, ancient, 55
Chemical system of medicine, 169
Cheselden, 216
Cheyne, John, 248
Chinese, medicine of the, 5
Chiron, 7
Chloroform, discovery of, 303
 Simpson's introduction of, 313
Chopart, 314
Circulation, capillary, discovery of, 158
 discovery of the, 155, 160
 lesser, 112
 failure to discover the, 113
Civiale, 269
Clark, Alonzo, 288
Classification of events in the history of medicine, 12
Clinical teaching, earliest systematic, 167
Cloquet, 269
Cnidus, Temple of Æsculapius at, 9
Cocaine, 314
Coction, doctrine of, 24
Colles, Abram, 248
Collot family of lithotomists, 177
Colonial physicians, 276
Columbus, 107, 155
Compass, invention of the, 99
Constantine the African, 74
Contrastimolo, 240
Cooper, Bransby, 273
 Samuel, 273
 Sir Astley, 271
Cornelius Agrippa, 139
Corpuscles of the blood, first discovery of, 158
Corvisart, 168
Cos, 19
 Temple of Æsculapius at, 9

Cosmogony, Greek, 13
Countess's powder, 165
Cowper, 158, 182, 183
Cox, John R., 286
Crisis, doctrine of, 25
Cruveilhier, 245
Cullen, William, 198
Currie, 229
Czermak, 253

Dalton, John C., 288
Darwin, Charles, 237
 Erasmus, 202
Daviel, 215
De Graaf, 183
De Haën, 200
De la Marche, Marguerite, 182
Delafield, Edward, 290
De Launay, 177
Delamotte, 166, 182
Delpech, 268
De Marque, 177
Denis, Jean Baptiste, 176
Denman, Thomas, 220
Dental surgery, the first college of, 341
Dentistry, ancient and mediæval, 332
 as a specialty of medicine, 337
 in America, 341
 of prehistoric times, 331
 relation of, to modern surgical pathology, 340
Desault, P. J., 214, 267
Devanter, 166
Dewees, William P., 288
Diagnosis, exact methods in, 263
Dionis, Pierre, 177
Dissection, ceremonials previous to, 149
 difficulties attending, 103
 of human bodies, 32
Doctor's mob in New York, 284
Dodart, 172

Eberle, John, 286
Eclectics, 14, 46
Embalming, 3
Empedocles, 17
Empirics, 14
Engraving, 100
Engravings, first anatomical, 112
Epidaurus, Temple of Æsculapius at, 10
Erasistratus, 34
Erudite period, 13
Esquirol, 228
Ether, sulphuric, as an anæsthetic, 302
Eustachius, 107
Eve, Paul F., 294

Fabre, 164
Fabricius ab Aquapendente, 109
Fabricius Hildanus, 110, 178

INDEX. 345

Fallopius, 109
Faust, 100
Fermentation, the causes of, 319
Fernel, Jean, 115
Fidelis, on legal medicine, 167
Filkin, 217
First hospitals in United States, 283
First medical schools in the United States, 281
Flint, Austin, 288
Fothergill, John, 212
Francis, John W., 286
Frank, J. P., 212
Frère Côme, 214
Frère Jacques, 177
Frick, 290
Fucus, 253
Functions of the spinal nerves, discovery of the, 248

Gaddesden, John, 87
Galen, anatomy and physiology of, 39
Galen, Claudius, 36
Galen's, influence, 43
 theories, 39
Gardiner, 199
Garengeot, 213
Gaub, 195
Gerard of Cremona, 90
Gerdy, 270
Germicides, internal use of, 329
Germ-theory of disease, 259
 what it means, 323
Gibson, William, 263
Gilbert, of England, 87
Gimbernat, 215
Glisson, 163, 183
Goerter, 163
Goode, John Mason, 247
Goodwin, 160
Goursaud, 177
Graves, Robert, 248
Gray, John P., 290
Greece, medicine of, 16
Greek period, 12, 49
Gregory, 199
Griesinger, 254
Gross, Samuel D., 294
Guillemeau, Jacob, 131
Gunn, Moses, 295
Guthrie, 273
Guttenberg, 100
Guy de Chauliac, 93
Gymnasia, the, 18
Gymnasiarch, 18
Gymnast, 18

Hahn, 229
Hahnemann, 241
Haller, 160, 162, 163, 220
Haly-Abbas, 61

Hamilton, Frank H., 293
Harvey, Gideon, 175
 William, 155
Hasheesh, 301
Hasner, 253
Havers, 183
Heberden, 212
Hebra, 253
Hebrews, medicine of the, 3
Heister, 215
Helvetius, 160
Henle, 254
Henri de Mondeville, 88
Herophilus, 33
Hessenfratz, 160
Heurne, Otto de, introduction of bedside instruction by, 167
Hewson, 161
Hiera sacra, 3
Highmore, 183
Hippocrates, 19
Hirudinomania, 244
Hoboken, 183
Hodgen, John T., 294
Hoffmann, Christopher Ludwig, 200
Hoffmann, Friedrich, 197
Hoffmann's dynamic system, 197
Holmes, Oliver Wendell, 290
Home, Sir Everard, 224
Homœopathy, 211
Horner, W. E., 287
Hospitals and clinics, 231
Howard, John, 212
Hufeland, 240
Humanization of vaccine-virus, 228
Hunter, John, 164, 218
 William, 218
Hunters' study of the lymphatic system, the, 161
Huxham, John, 212
Hydrotherapeutic system, 255
Hydrotherapy, 229

Iatrochemical system, 169
Iatroliptes, 18
Iatromechanical school, 171
Infarctus, doctrine of, 201
Influence, of botany on medicine, 237
 of chemistry on medicine, 238
 of Darwin and Spencer on medicine, 237
 of Harvey's discovery, 159
 of physics on medicine, 238
 of the art of printing, 100
 of the French Revolution, 191
 of the Northern invaders, 71
 of the occult sciences, 139
 of the Salernian school, 81
 of zoölogy on medicine, 238
Inoculation, against small-pox, in America, 279

Inoculation, preventive, against small-pox, 225
　with cow-pox, the first, 226
Irritability of tissues, discovery of, 163
Isopathy, 241

Jackson, Charles T., 310
Jacobus Sylvius, 103
Jäger, 253
Jaxtthal, 253
Jenner, Edward, 226
Jesuit powder, 165
Jews, prejudice against the, 233
John Actuarius, 66
John of Procida, 79
Joubert's *Popular Errors*, 147

Kampf, 201
Kepler, 162
Keyes, 289
Kirkbride, 290

Laennec, 262
Lamballe, Joubert de, 270
Lancisi's clinic, 168
Lanfranc, 91,
Langenbeck, Bernhard von, 264
Larrey, 267
Lavoisier, 160, 191
Lawrence, Sir William, 274
Le Boe, 167, 169
Le Cat, 214
Le Dran, 213
Léoniceuus, Nicholas, 101
Lettsom, 213
Leuwenhoek, 158
Levret's modification of obstetrical forceps, 166
Ligatures, first use of, in amputations, 127
Linacre, Thomas, 101
Linnæus, 191
Lisfranc, 269
Lister, 261
Lister's, studies and methods, 325
　work, benefits of, 327
Liston, Robert, 274
Lithotomy, lateral, inventor of, 177
Lizars, John, 274
Long, Crawford, 304
Lorry, 228
Louis, 246
Lymph, discovery of the circulation of, 158

Machaon, 10
McClellan, George, 293
McDowell, Ephraim, 267, 292
Magati, 176
Magendie, 246
Magic, 141
Magnetism, animal, 203

Maimonides, 65
Maitre, Jean, 178
Malgaigne, 270
Malpighi, 158
Mandragora, 301
Marcellus Donatus, 115
March, Alden, 295
Marchetti, 158, 176
Mareschal, Georges, 178
Marinus, 42
Marjolin, 269
Mascagni, 161
Mauriceau, 166, 182
Mauthner, 253
Mayow, 160
Mead, Richard, 213
Mechanico-dynamic system of medicine, 197
Meckel, 162
Medical, journals in the United States, first, 285
　jurisprudence, beginning of, 166
　school of the natural sciences, 258
　study under preceptors, 277
　systems, promulgation of, 152
Medici puri, pretensions of, 189
Medicine, and surgery, approach of, 147
　dogmatic school of, 13
　Imperial school of, at Pekin, 6
　physiological theory of, 243
　separation of priesthood from, 147
Meibom, 184
Meigs, John Forsyth, 289
Melampus, 6
Mesmer, Frank, 203
Mesmerism, 203
Mesue, 60
Methodism, 13, 45
Microscope, 100
Midwifery during the seventeenth century, advance of, 182
Midwives, 165
Mondino, 92
Monro, Alexander, Sr., 216
　Alexander, second and third, 216
　Donald, 216
Monroes, the two, 164
Montpellier, the school of, 86
Morand, 213
Morel, 176
Morgagni, 224
Morton, Richard, 175
　William T. G., 306
Mott, Valentine, 293
Müller, 222
Munich Clinical School, 258
Muralt, 180
Mutter, Thomas D., 293
Mystic period, 12

Natural history, the school of, 249

Natural philosophy, the school of, 249
Needham, 183
Nélaton, 270
Nepenthe, 301
New Vienna School, 250
Nitrous-oxide gas, 303
Nominalist, 69
Nuck, 183

Obstetrical forceps, invention of the, 166
Obstetricians and gynæcologists, American, 295
Obstetrics, development of, 166
Oken, 249
Ophthalmoscope, 263
Oppolzer, 253
Oribasius, 48
Orthopædics, origin of name, 177
Ovariotomy, the first, 267

Pacchioni, 183
Palfyn's obstetrical forceps, 166
Paracelsus, 143
Paré, Ambroise, 123
Paré and the surgeons of St. Côme, 131
Park, Henry, 217
Parker, Willard, 293
Paulus Ægineta, 51
Pecquet, 161
Percussion, invention of the art of, 210
Pergamos, library of, 30
Periodic physicians, 17
Peruvian bark, discovery of, 164
Petit, J. L., 213
Pen, 166
Peyer, 183
Peyronie, 213
Pfeufer, 254
Pharmacopolists, 54
Philosophic period, 12, 18
Phrenology, 163, 242
Physical examination, methods of, 263
Physick, Philip S., 291
Physiological medicine, 253
Pinel, 163, 196, 206, 223
Piorry, 262
Pitard, John, 92
Plater, Felix, 118
Plato, 27
Pleximeter, 263
Podalirius, 10
Porta, Giovanni Batista, 118
Portal, 223
    Paul, 166, 182
Pott, Percival, 217
Pravaz, 269
Praxagoras, 27
Priessnitz, 255

Primitive period, 12
Pringle, Sir John, 212
Ptolemy Soter, 31
Purkinje, 222
Purmann, 180
Pythagoras, 15

Quesnay, 173, 214
Quintus, 42

Radcliffe, 213
Rademacher, 254
Rapid multiplication of scientific literature, 239
Rasori, 240
Rau, 180
Ray, Isaac, 290
Raymond Lulli, 87
Realism, 206
Realist, 69
Receptaculum chyli, discovery of, 161
Reflex action, discovery of, 248
Reform period, 13
Regulation of practice in colonial times, 285
Reil, 202
Religious orders and the sick, 95
Rembert Dodoens, 115
Reuss, 253
Rhazes, 60
Rhinoplasty, 176
Rhodes, Temple of Æsculapius at, 8
Richerand, 267
Richter, August Gottlieb, 216
Riolan, 128
Rivinius, 183
Rodger, J. K., 293
Roeschlaub, 240
Roger of Parma, 78
Roland of Parma, 78
Rolfink, 188
Rome during the Greek period, 53
Rokitansky, 250
Roser, 253
Rousset and the Cæsarean operation, 134
Roux, 269
Rufus of Ephesus, 42
Rush, Benjamin, 206, 283
Ruysch, 158

Sabatier, 214
Sacred period, 12
Salernum, school of, 72
Sandifort, 219
Sands, Henry B., 293
Sanson, 269
Santoro, 171
Santoro's thermometer, **171**
Sauvage, 196
Saviard, 177

348 INDEX.

Scalpel, first use of, in dissecting, 112
Scarpa, 162, 215
Schaffhausen, 183
Schneider, 162, 184
School of rational medicine, 254
Scientific societies and journals, origin of, 154
Scultetus, 180
Seminalism, 259
Serapion, 60
Servetus, Michael, 112, 155
Severino, 119, 176
Shoeffer, 100
Shot wounds, the new teaching of Paré concerning, 132
Siegemundin, Justine, 182
Sigmund, 253
Simpson, Sir James Y., 274
Sims, J. Marion, 296
Skoda, 251
Smellie, William, 220
Smellie's modification of the obstetrical forceps, 166
Smith, Nathan R., 292
Societies and academies, foundation of, 235
Soemmering, 162, 222
Solidism, 194
Spontaneous generation of life disproven, 321
St. Côme, College of, 92, 122
Stahl, 195
Stahl's pietistic system, 195
Steno, Nicholas, 159
Steno's duct, 159
Stethoscope, 262
Stimolo and contrastimolo, 240
Stoerck, 200
Stokes, William, 248
Stoll, 200
Student-life during the fifteenth and sixteenth centuries, 148
Surgery, achievements of, 263
reasons for neglect of, 120
Swammerdam, 153, 183
Sydenham, 152, 165, 173
Sylvius, 167
Syme, James, 274
Syphilis, wide-spread outbreak of, 136

Teeth, substitutes for human, 335
Telescope, invention of the, 99
Tenon, 215
Thaer, 199
Themison, 44
Theory of excitement, 246
Theosophy, 141
Thermometer, discovery of the, 171
Thoth, 2
Tourniquet, invention of, 176
Tourniquet, screw, invention of, 213

Transfusion of blood in man, the first, 176
Travers, Benjamin, 248, 273
Treatment of the insane, improvement in, 228
Troja, 228
Trotula, 79
Trousseau, 247
Türck, 253
Tyrrel, 273

Universities and royal scientific societies, 192

Vaccination, compulsory, 228
in the United States, the first, 279
the first, 227
Vagadasastir, 4
Valsalva, 176
Van Buren, William H., 289
Van Helmont, 168
Van Helmont's system of medicine, 168
Van Siebold, 216
Van Swieten, 168
Van Swieten and the Old Vienna School, 199
Velpeau, 270
Venesection, first account of, 10
revival of, 118
Verulam, Lord, 153
Vesalius, Andreas, 104
Vicq d'Azyr, 162, 164, 223
Vidal, 269
Vieussens, 162, 186
Virchow, Rudolph, 255
Vitalism, 201
modern, 255

Wainman, 217
Warren, John Collins, 291
Waterhouse, Dr., 279
Wells, Horace, 305
Werlhof, 211
Wharton, 183
White, Anthony, 216
Charles, 216
Wichman, 211
William of Salicet, 91
Willis, Thomas, 163, 170
Winslow, 164, 223
Winternitz, 255
Wirsung, 183
Wiseman, Richard, 180
Wistar, Caspar, 286
Wolf, 222
Wood, George B., 287
Wren, Sir Christopher, 182
Wunderlich, 250, 254

Zeissel, 253

SEE NOTE AT FOOT OF PAGE 4.

Brief Catalogue of the Publications

OF

THE F. A. DAVIS CO.

PHILADELPHIA : 1914-16 Cherry Street.
NEW YORK : 117 West Forty-Second Street.
CHICAGO : 9 Lakeside Building, 214-220 South Clark Street.

WRITE FOR OUR COMPLETE DESCRIPTIVE CATALOGUE.

BASHORE, Harvey B., M.D. Improved Clinical Chart for the Separate Plotting of Temperature, Pulse, and Respiration. But one color of ink necessary. Designed for the Convenient, Accurate, and Permanent Daily Recording of Cases in Hospital and Private Practice. Fifty Charts, in Tablet Form. Size, 8 x 12 inches . . . . Net, $0.50

BASHORE, Harvey B., M.D. Outlines of Rural Hygiene for Physicians, Students, and Sanitarians, with an Appendix on "The Normal Distribution of Chlorine," by Herbert E. Smith, M.D. Illustrated. 12mo. About 100 pages. Extra Cloth . . . . Net, .75

BAUDRY, S., M.D. (Lille, France). Injuries to the Eye in their Medicolegal Aspect. Translated with the permission of the author and edited by Charles A. Oliver, A.M., M.D. *Ready Shortly.* Crown Octavo. About 200 pages. Extra Cloth . . . . Net, 1.25

BISHOP, S. S., M.D., LL.D. Diseases of the Ear, Nose, and Throat. A Comprehensive Practical Text-book. Thoroughly Illustrated with numerous Engravings and several Full-page Colored Plates. Royal Octavo. Over 500 pages. Extra Cloth, Net, $4.00. Sheep or Half-Russia . . . . Net, 5.00

BOENNING, Henry C., M.D. Textbook on Practical Anatomy. Including a Section on Surgical Anatomy. About 200 Wood-engravings. Royal Octavo. Nearly 500 pages. Extra Cloth. Also in Oil-cloth, for use in the dissecting-room without soiling . . . . Net, 2.50

BOUCHARD, Prof. Ch. (Paris). Auto-Intoxication. Being a Series of Lectures on Intestinal and Urinary Pathology. Translated from the French, with an Original Appendix, by Thomas Oliver, M.A., M.D. Over 300 pages. Crown Octavo. Extra Cloth . . . . . . Net, 1.75

BOWEN, Cuthbert, M.D. Hand-book of Materia Medica, Pharmacy, and Therapeutics. 12mo. 370 pages. Extra Cloth . . . . Net, 1.40

BURET, Dr. F. (Paris). Syphilis To-day and Among the Ancients. In three volumes. Translated from the French, with notes, by A. H. Ohmann-Dumesnil, M.D.
Volume I, Syphilis in Ancient and Prehistoric Times, with a Chapter on the Rational Treatment of Syphilis in the Nineteenth Century. 280 pages. 12mo. Cloth . . . . Net, 1.25
Volumes II and III, bound in one, Syphilis in the Middle Ages and Syphilis in Modern Times. 12mo. 300 pages. Cloth . . . . Net, 1.50

CAPP, William M., M.D. The Daughter: Her Health, Education, and Wedlock. Homely Suggestions to Mothers and Daughters. 12mo. 150 pages. Cloth, Net, $1.00. In Paper Covers (unabridged) . . . . Net, $0.50

CATHELL, D. W., M.D. Book on the Physician Himself and Things that Concern his Reputation and Success. Tenth Edition. Author's last revision. Royal Octavo. 350 pages. Cloth . . . . Net, 2.00

CLEVENGER, S. V., M.D. Spinal Concussion. Royal Octavo. Nearly 400 pages. With 30 Wood-engravings. Net, 2.50

COLTMAN, Robert, Jr., M.D. The Chinese: Their Present and Future. Medical, Political, and Social. Fifteen Fine Engravings on Extra Plate Paper, from photographs of persons, places, and objects characteristic of China. Royal Octavo. 212 pages. Extra Cloth, with Chinese Side-stamp in gold . . . Net, 1.75

CROTHERS, Geo. D., A.M., M.D., and RICE, Hiram H., A.M. Elements of Latin for Students of Medicine and Pharmacy. Neat 12mo. Flexible Binding . . . . Net, 1.25

DAVIS, N. S., Jr., A.M., M.D. Consumption : How to Prevent it and How to Live with it. Its Nature, Causes, Prevention, and the Mode of Life, Climate, Exercise, Food, and Clothing Necessary for its Cure. 12mo. 143 pages. Cloth . . . . Net, .75

DAVIS, N. S., Jr., A.M., M.D. Diseases of the Lungs, Heart, and Kidneys. 12mo. Over 300 pages. Cloth . . . Net, 1.25

DEMARQUAY, J. N., M.D. (Paris). On Oxygen. A Practical Investigation of the Clinical and Therapeutic Value of the Gases in Medical and Surgical Practice, with Especial Reference to the Value and Availability of Oxygen, Nitrogen, Hydrogen, and Nitrogen Monoxide. Translated with notes, additions, and omissions, by Samuel S. Wallian, A.M., M.D. Royal Octavo. 316 pages. Illustrated with 21 Wood-cuts. Net, $2.00. Half-Russia . . . . Net, 3.00

EDINGER, Dr. Ludwig, Structure of the Central Nervous System. Second, Revised American Edition. From the Last German Edition. With 150 Illustrations. Translated by W. H. Vittum, M.D. Edited by C. E. Riggs, A.M., M.D. Royal Octavo. About 500 pages. Extra Cloth . . . . Net, 1.75

EISENBERG, James, Ph.D., M.D., Vienna. Bacteriological Diagnosis. Tabular Aids for Use in Practical Work.

Translated and Augmented from the Second German Edition, by Norval H. Pierce, M.D. Nearly 200 pages. Royal Octavo, bound in Cloth and in Oil-cloth (for laboratory use) . . . . Net, $1.50

ESHNER, Augustus A., M.D. Handbook of Fevers, Including General Considerations, Typhoid Fever, Typhus Fever, Influenza, Malarial Fever, Yellow Fever, Variola, Relapsing Fever, Weil's Disease, Thermic Fever, Dengue, Miliary Fever, Mountain Fever, etc.; their Prevention, Etiology, Pathology, Diagnosis, Prognosis, and Treatment. Compiled from The Annual of the Universal Medical Sciences, from 1888 to 1894, inclusive, with copious commentaries and additions. Illustrated by Wood-engravings and Lithographs. Royal Octavo. Over 600 pages. Cloth, Net 3.50

FIREBAUGH, Ellen M. The Physician's Wife and the Things that Pertain to Her Life. Crown Octavo. 200 pages, with 44 Original Character Illustrations and a Frontispiece Portrait of the Author. Extra Cloth . . . . Net, 1.25
Special Limited Edition. First 500 Copies printed in Photogravure Ink on Extra-Quality Enameled Paper, with wide margins. Bound in Fine Vellum Cloth and Leather . . . . Net, 3.00

GANT, S. G., M.D. Diagnosis and Treatment of Diseases of the Rectum, Anus, and Contiguous Textures. Designed for Practitioners and Students. With two Chapters on Cancer and Colotomy by Herbert Wm. Allingham, F.R.C.S.Eng. Illustrated with 16 Full-page Chromolithographic Plates and about 125 Photo-engravings in the Text. Royal Octavo. Over 400 pages. Extra Cloth, Net, $3.50. Half-Russia, Gilt Top, Net 4.50

GOODELL, Wm., A.M., M.D., etc. Lessons in Gynecology. With 112 Illustrations. Third Edition, thoroughly revised and greatly enlarged. One volume. Large Octavo. 578 pages. Cloth, $5; Full Sheep, $6. Discount, 20 per cent., making it, net, Cloth, $4; Sheep, $4.80. Postage 27 cents extra.

GRANDIN, Egbert H., M.D., and JARMAN, George W., M.D. Practical Obstetrics, embracing Pregnancy, Labor, the Puerperal State, and Obstetric Surgery. A Text-book for Physicians and Students. Royal Octavo. Over 500 pages, with more than 50 Full-page Photogravure Plates taken from Nature, besides many other cuts in the text. Cloth, net, $4.00. Sheep . . Net, 4.75

GUERNSEY, Henry N., M.D. Plain Talks on Avoided Subjects. 16mo. Extra Cloth 1.00

HARE, Hobart Amory, M.D., B.Sc. Epilepsy: Its Pathology and Treatment. 12mo. 228 pages. Cloth . . . . Net, 1.25

HARE, Hobart Amory, M.D., B.Sc. Fever: Its Pathology and Treatment. Containing Directions and the Latest Information Concerning the Use of the So-called Antipyretics in Fever and Pain. Illustrated with more than 25 new plates of tracings of various fever cases. 12mo. Cloth . . . . Net, 1.25

HUIDEKOPER, R. S., M.D. Age of the Domestic Animals. Being a Complete Treatise on the Dentition of the Horse, Ox, Sheep, Hog, and Dog, and on the various other means of determining the age of these animals. Royal Octavo. 225 pages. 200 Wood-engravings. Extra Cloth . . . . Net, $1.75

INTERNATIONAL SYSTEM OF ELECTRO-THERAPEUTICS. For Students, General Practitioners, and Specialists. Chief Editor, Horatio R. Bigelow, M.D. Assisted by Thirty-eight Eminent Specialists in Europe and America as Associate Editors. Thoroughly Illustrated with many fine Engravings. 1160 pages. Royal Octavo. Extra Cloth, net, $6.00. Sheep, net, $7.00. Half-Russia . . . . . . Net, 7.50

INTERNATIONAL TEXT-BOOK OF MEDICAL ELECTRO-PHYSICS AND GALVANISM. For the Use of Medical Students and Practitioners. Being the first nine Chapters or Articles of the International System of Electro-Therapeutics. By W. J. Herdman, Ph. B., M.D.; H. McClure, M.D.; J. M. Bleyer, M.D.; W. F. Robinson, M.D.; A. W. Duff, M.A., B.Sc. (Ed.); G. J. Engelmann, M.D.; A. P. Brubaker, M.D.; F. Peterson, M.D; W. Mills, M.A., M.D., L.R.C.P. (Lond.), F.R.S. (Can.). Thoroughly Illustrated. Royal Octavo. About 400 pages. Cloth . . . . . Net, 2.50

IVINS, Horace F., M.D. Diseases of the Nose and Throat. Royal Octavo. 507 pages. 129 Illustrations. Extra Cloth, net, $4.00. Sheep or Half-Russia . . . . Net, 5.00

JENNINGS, J. Ellis, M.D. Color-Vision and Color-Blindness. A Practical Manual for Railroad-Surgeons. Crown Octavo. Over 100 pages. 21 Engravings and 1 Full-page Colored Plate. Cloth. Net 1.00

KEATING, John M., M.D. Record-book of Medical Examinations for Life-insurance. Covering one hundred examinations. Size, 7 x 3½ inches. Cloth, Net 50

KRAFFT-EBING, Dr. R. von. A Text-book on Insanity. Authorized translation of the Fifth German Edition by C. G. Chaddock, M.D. Royal Octavo. About 800 pages. *In Preparation.*

LIEBIG, G. A., Jr., Ph.D., and ROHE, George H., M.D. Electricity in Medicine and Surgery. Profusely Illustrated. Royal Octavo. 383 pages. Extra Cloth. Net 2.00

MANTON, Walter Porter, M.D. A Syllabus of Lectures on Human Embryology. An Introduction to the Study of Obstetrics and Gynæcology, with a Glossary of Embryological Terms. Second (Revised) Edition. Interleaved for Taking notes, and Thoroughly Illustrated. 12mo. Extra Cloth . . . . Net, 1.25

MASSEY, G. Betton, M.D. Electricity in the Medical and Surgical Diseases of Women. With Special Reference to the Apostoli Treatment of Fibroid Tumors and the Cataphoric Treatment of Malignant Disease. Third Edition. Revised and Enlarged. With New and Original Wood-engravings. Also several Full-page Photographic Plates from Nature. *In Press. Nearly Ready.*

MEDICAL BULLETIN VISITING-LIST, OR PHYSICIANS' CALL-RECORD. Arranged upon an Original and Convenient Monthly and Weekly Plan for the Daily Recording of Professional Visits. Handsomely bound in fine, strong leather. Size 4 x 6¾ inches. In three styles. *Send for descriptive circular.*

No. 1. For 70 patients daily each month for one year . . . . Net, $1.25

No. 2. For 105 patients daily each month for one year . . . . Net, 1.50

No. 3. In which "The Blanks for Recording Visits in" are in six removable sections and adapted to the smallest or largest possible practice . . . . . . Net, 1.75

MICHENER. Hand-book of Eclampsia, or Notes and Cases of Puerperal Convulsions. By E. Michener, M.D.; J. H. Stubbs, M.D.; R. B. Ewing, M.D.; E. Thompson, M.D.; S. Stebbins, M.D. 16mo, Cloth . . . . Net, .60

MONTGOMERY, E. E., M.D. Practical Gynæcology. A Text-book for Students and Physicians. Royal Octavo. Thoroughly Illustrated. *In Preparation.*

NISSEN, Prof. Hartvig. A Manual of Instruction for Giving Swedish Movement and Massage Treatment. With 29 Original Wood-engravings. 12mo. 128 pages. Cloth . . . . Net, 1.00

PARK, Roswell, A.M., M.D. An Epitome of the History of Medicine from the Earliest Times to the Present. Handsomely Illustrated with Half-tone Reproductions of Portraits of the Eminent Physicians and Surgeons of all time, and other Engravings. Crown Octavo. 350 Pages. *Extra Cloth. Beveled Edges.* Net 2.00

PHYSICIANS' INTERPRETER. In English, French, German, and Italian. Specially arranged for diagnosis by M. you V. Full Russia Leather, for carrying in the pocket. Size, 5 x 2⅔ inches. 206 pages . . . . Net, 1.00

PHYSICIANS' ALL - REQUISITE TIME- AND LABOR-SAVING ACCOUNT-BOOK. Being a Ledger and Account-Book for Physicians' Use, meeting all the Requirements of the Law and Courts. Designed by Wm. A. Seibert, M.D.

No. 1. 300 pages for 900 Accounts per Year, size 10 x 12, bound in ¾-Russia. Raised Back-bands. Cloth Sides . . Net, 5.00

No. 2. 600 pages for 1800 Accounts per Year, size 10 x 12, bound in ¾-Russia. Raised Back-bands, Cloth Sides . . Net, 8.00

Send for Complete Descriptive Circular.

PRICE, W. Henry, M.D., and EAGLETON, S. Potts, M.D. Three Charts of the Nervo-Vascular System.

Part I. The Nerves. Part II. The Arteries. Part III. The Veins. Each Chart 19 x 24 inches.

New Edition. Printed upon Extra-durable Paper. Per set . . . . . . Net, .50

PURDY, Chas. W., M.D. Practical Uranalysis and Urinary Diagnosis. Third (Revised) Edition. With numerous Illustrations, including several Colored Plates. Crown Octavo. Nearly 400 pages. Extra Cloth . . . . . . Net, 2.50

PURDY, Chas. W., M.D. Diabetes: Its Cause, Symptoms, and Treatment. 12mo. 184 pages. Cloth . . . . . Net, 1.25

RANNEY, A. L., A.M., M.D. Eye-strain in Health and Disease. With Special Reference to the Amelioration or Cure of Chronic Nervous Derangements Without the aid of Drugs. A Manual for the General Practitioner in Medicine to aid in Detecting the Presence and Appreciating the Importance of Eye-strain in Health and Disease. Fully Illustrated. Royal Octavo. 325 pages. *Extra Cloth. Beveled Edges.* Net 2.00

REMONDINO, P. C., M.D. History of Circumcision from the Earliest Times to the Present. Moral and Physical Reasons for its Performance: with a History of Eunuchism, Hermaphrodism, etc., and of the Different Operations Practiced on the Prepuce. 12mo. 346 pages. Cloth . . . . Net, $1.25

REMONDINO, P. C., M.D. The Mediterranean Shores of America. Southern California: Its Climate, Physical, and Meteorological Conditions. Royal Octavo. 175 pages. With 45 Illustrations and 2 Maps of this region, showing altitudes, ocean-currents, etc. Cloth . Net, 1.25

Cheaper edition (unabridged) bound in Paper . . . . Net, .75

ROHÉ, George H., M.D. Text-book of Hygiene. Third Edition. Carefully Revised and Enlarged, with many Illustrations and valuable Tables. Royal Octavo. Over 450 pages. Extra Cloth. Net 3.00

ROHÉ, George H., M.D. A Practical Manual of Diseases of the Skin. By George H. Rohé, M.D., assisted by J. Williams Lord, A.B., M.D. Appropriately Illustrated. *New Edition Preparing.*

SAJOUS, Charles E., M.D. Hay Fever and its Successful Treatment by Superficial Organic Alteration of the Nasal Mucous Membrane. With 13 Engravings on Wood. 12mo. Extra Cloth. Net 1.00

SENN, N., M.D., Ph.D., LL.D. Principles of Surgery. New (Second) Edition Thoroughly Revised and Considerably Enlarged. Royal Octavo. 180 Wood-engravings and several full-page Colored Plates. 660 pages. Cloth, net, $4.50. Sheep or Half-Russia . . . Net, 5.50

SENN, N., M.D., Ph.D., LL.D. Tuberculosis of the Bones and Joints. Royal Octavo. Over 500 pages. Illustrated with 107 Engravings. Extra Cloth, net, $4.00. Sheep or Half-Russia . . . Net, 5.00

SHOEMAKER, John V., M.D., LL.D. Heredity, Health, and Personal Beauty, Including the Selection of the best Cosmetics for the Skin, Hair, Nails, and All Parts Relating to the Body. Royal Octavo. 425 pages. Cloth, net, $2.50. Half-Morocco . . . . Net, 3.50

SHOEMAKER, John V., M.D., LL.D. Materia Medica and Therapeutics. A Practical Text-book for Students and Practitioners. *Fourth Edition.* Thoroughly Revised and in Conformity with the Latest Revision of the United States Pharmacopœia. Royal Octavo. 1100 pages. Cloth, net, $5.00. Sheep . Net, 5.75

SMITH, Robert Meade, A.M., M.D. Physiology of the Domestic Animals. Royal Octavo. Over 950 pages. Profusely Illustrated. Cloth, net, $5.00. Sheep . . . . Net, 6.00

SOZINSKEY, Thomas S., M.D., Ph.D. Medical Symbolism. Historical Studies in the Arts of Healing and Hygiene. Illustrated with 30 new Wood-engravings. 12mo. Nearly 200 pages. Cloth. Net 1.00

STEWART, John S., M.D. Obstetric Synopsis. A Complete Compend. With an Introductory Note by William S. Stewart, A.M., M.D. 42 Illustrations. 202 pages. 12mo. Cloth . . . . . Net, 1.00

STRAUB, D. W., M.D. Symptom Register and Case Record. Published in stiff Board Tablets, of 50 sheets each, at 50 cents, net, per tablet, and in Book form, flexible binding, with Alphabetical Marginal Index . . . . Net, $0.75

VOUGHT, Walter, Ph.B., M.D. Chapter on Cholera for Lay Readers. History, Symptoms, Prevention, and Treatment of the Disease. Illustrated. 12mo. 106 pages. Flexible Cloth 75

WITHERSTINE, C. Sumner, M.S., M.D. International Pocket Medical Formulary. Arranged Therapeutically. Including more than 1800 formulæ from several hundred well-known authorities. With a valuable Appendix. 275 printed pages, besides extra blank leaves for new formulæ. Bound in Leather, with Sideflap . . . . Net, $2.00

YOUNG, James K., M.D. Synopsis of Human Anatomy. Being a Complete Compend of Anatomy, including the Anatomy of the Viscera, and numerous Tables. Illustrated with 76 Wood-engravings. 320 pages. 12mo. Cloth. Net 1.00

The following Publications are Sold only by Subscription.

SAJOUS'S ANNUAL AND ANALYTICAL CYCLOPÆDIA OF PRACTICAL MEDICINE. In Six Royal Octavo Volumes. Thoroughly illustrated with Wood-engravings and Colored Plates. Volume I ready in November, 1897. Other volumes to follow at intervals of about four months. Price, per volume, in Cloth, net, $5.00. Half-Russia . . . . Net, $6.00

ANNUAL OF THE UNIVERSAL MEDICAL SCIENCES: The Standard Medical Annual of the World. A Yearly Report of the Progress of the General Sanitary Sciences Throughout the World. Edited by Charles E. Sajous, M.D., and Seventy Associate Editors, assisted by over Two Hundred Corresponding Editors and Collaborators. In Five Royal Octavo Volumes of about 500 pages each. Illustrated with Chromolithographs, Engravings, Maps, Charts, and Diagrams. Being intended to enable any physician to possess, at a moderate cost, a complete Contemporary History of Universal Medicine. Subscription Price per year (including the *Universal Medical Journal,* issued monthly, for one year), Cloth, 5 vols., Royal Octavo, $15.00. Half-Russia . . . . . . Net, 20.00

ADAMS, J. Howe, M.D. History of the Life of D. Hayes Agnew, M.D., LL.D. A fascinating life-history of one of the world's greatest surgeons. Royal Octavo, 376 pages. Handsomely printed, with Portraits and other Illustrations. Extra Cloth, net, $2.50. Half-Morocco, Gilt Top . . . . Net, 3.50

KRAFFT-EBING, Dr. R. von. Psychopathia Sexualis. With Especial Reference to Contrary Sexual Instinct. A Medico-legal Study of Sexual Insanity. Authorized Translation of the Seventh German Edition, by C. G. Chaddock, M.D. Royal Octavo. 432 pages. Cloth, net, $3.00. Sheep . . . . Net, 4.00

RANNEY, Ambrose L., A.M., M.D. Lectures on Nervous Diseases. From the Stand-point of Cerebral and Spinal Localization, and the Later Methods Employed in the Diagnosis and Treatment of these Affections. Profusely Illustrated with Original Diagrams and Sketches in Color by the author, carefully selected Wood-engravings, and Reproduced Photographs of Typical Cases. Royal Octavo. 780 pages. Cloth, $5.50, Sheep, $6.50. Half-Russia . $7.00

SAJOUS, Charles E., M.D. Lectures on the Diseases of the Nose and Throat. Delivered at the Jefferson Medical College, Philadelphia. Illustrated with 100 Chromolithographs, from Oil-paintings by the author, and 93 Engravings on Wood. Royal Octavo. Cloth, $4.00. Half-Russia 5.00

SCHRENCK-NOTZING, Dr. A. von. Suggestive Therapeutics in Psychopathia Sexualis. Authorized Translation of the Latest German Edition, by C. G. Chaddock, M.D. A supplementary volume to Dr. R. von Krafft-Ebing's treatise on "Psychopathia Sexualis" (also translated by Dr. Chaddock). Royal Octavo. About 350 pages. Cloth, net, $2.50. Sheep . . . . Net, 3.50

STANTON, Mary Olmstead. The Encyclopedia of Face and Form Reading. Or Personal Traits, both Physical and Mental, Revealed by Outward Signs Through Practical and Scientific Physiognomy. With an outline of study, glossary, and classified suggestive questions and elaborate aids to the study, together with original articles upon vital subjects by distinguished authorities. Second Edition, Revised. Profusely Illustrated. Royal Octavo, over 1300 pages. Cloth, $4.00; Sheep or Half-Russia 5.00

JOURNAL OF LARYNGOLOGY, RHINOLOGY, AND OTOLOGY. A Monthly Analytical Record of Current Literature Relating to the Throat, Nose, and Ear. Edited by Dr. Norris Wolfenden, of London; Dr. John Macintyre, of Glasgow, and Dr. Dundas Grant. Price, per year, inclusive of postage 5.00

THE MEDICAL BULLETIN. A Monthly Journal of Medicine and Surgery. Edited by John V. Shoemaker, M.D., LL.D. Terms, per year, in advance, in United States, Canada, and Mexico 1.00

THE UNIVERSAL MEDICAL JOURNAL. A Monthly Magazine of the Progress of Every Branch of Medicine in All Parts of the World. Edited by Charles E. Sajous, M.D. Subscription Price, in United States, per year 2.00

NOTE.—Prices of all books in this list are absolutely net (excepting Guernsey's "Plain Talks" and Goodell's "Gynecology," which are subject to 20 per cent. discount) and no discount can be allowed retail purchasers therefrom, but all carriage charges will be prepaid by the publishers. All purchasers in Canada must pay Custom duties in addition to the foregoing prices.

www.ingramcontent.com/pod-product-compliance
Lightning Source LLC
Chambersburg PA
CBHW020316240426
43673CB00039B/824